D0456923

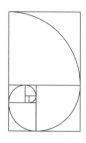

Mathematics

Dr. Keith Devlin has been a professional research mathematician since 1971, when he obtained his Ph.D. in mathematics from the University of Bristol. He has held positions at universities in England, Scotland, Norway, Germany, Canada, and the United States. Since 1987 he has resided in the United States and is currently dean of science and professor of mathematics at Saint Mary's College of California, in Moraga, California. He also is a senior researcher at Stanford University and a consulting research professor at the University of Pittsburgh.

Dr. Devlin is the author of twenty-three books on subjects in mathematics and computing, and he has published more than sixty research articles. Since 1983 he has written regular articles on mathematics and computing for *The Guardian* newspaper. A compilation of many of these articles was published in book form by the Mathematical Association of America in 1994 under the title *All the Math That's Fit to Print.* He served for six years as editor of *FOCUS,* the news magazine of the Mathematical Association of America, and has been a contributing editor of the *Notices of the American Mathematical Society.* Dr. Devlin is the author of "Devlin's Angle," a popular monthly mathematics column in the electronic journal *MAA Online* (www.maa.org). He appears regularly on National Public Radio in the United States and BBC Radio in the United Kingdom, commenting on mathematics and related issues, and he was a lead adviser for the six-part PBS television series *Life by the Numbers,* which was first broadcast in 1998. Dr. Devlin also wrote the official companion book to the series, under the same title, published by John Wiley and Sons.

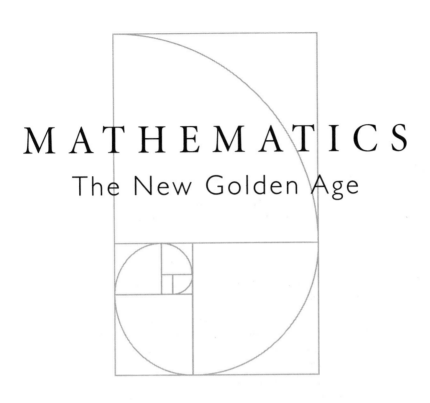

MATHEMATICS
The New Golden Age

Keith Devlin

COLUMBIA UNIVERSITY PRESS

NEW YORK

Columbia University Press
Publishers Since 1893
New York   Chichester, West Sussex

For permission to reproduce copyrighted material, grateful acknowledgment is made to the following publishers:

Figures 8 and 17 through 23 are reproduced from *The Beauty of Fractals* (1986), by H. O. Peitgen and P. H. Richter, with the permission of Springer-Verlag.

Figures 9 and 10 are reproduced from *Fractals: Form, Chance and Dimension* (1977), by Benoit Mandelbrot, with the permission of W. H. Freeman and Co.

Figure 14 is reproduced from *Studies in Geometry* (1970), by L. M. Blumenthal and K. Menger, with the permission of W. H. Freeman and Co.

Figures 31 and 34 are reproduced from *Scientific American,* with the permission of W. H. Freeman and Co.

Figure 57 is reproduced with the permission of Cordon Art bv.

Library of Congress Cataloging-in-Publication Data
Devlin, Keith J.
Mathematics : the new golden age / Keith Devlin.
p.   cm.
Includes index.
ISBN 0-231-11638-1 (cloth)
1. Mathematics Popular works.   2. Mathematics—History—20th century.   I. Title.
QA93.D46   1999
510—dc21   99-23438

∞

Contents

Preface

The new golden age of mathematics. When was the old one? Was it the period of the ancient Greek geometers around 300 B.C.? Or did it occur during the seventeenth century, when Newton and Leibniz were developing the infinitesimal calculus and Fermat was working on number theory? Or perhaps the mathematical career of Gauss alone (1777–1855) justifies the title *golden age.* Or was it still later, the period that saw the work of Riemann, Poincaré, Hilbert, and others? For the mathematics produced between the mid-1800s and the start of World War II was truly prodigious.

As with any area of human endeavor, it is impossible to say which was the truly "greatest period." Each new generation builds on the work of its predecessors. What we can say, however, is that the present time is witnessing a phenomenal amount of mathematical research. The 1998 edition of the *World Directory of Mathematicians* lists some 53,911 professional mathematicians* around the world, but this represents only a small fraction of the real total. If you include also the vast armies of "amateur mathematicians" (some of whom have made some significant discoveries) for whom mathematics is simply a pleasant pastime, then the true figure must be enormous. On grounds of numbers (admittedly

---

*The dictionary defines a professional mathematician as someone who has published at least two papers that have been reviewed by a professional organization.

shaky grounds, since quantity and quality are not closely related, especially in mathematics), we are in the middle of a new golden age right now. And since every book must have a title, that is more or less what I have chosen to call this one.

What I hope to do in this book is convey to the interested layperson some of the most significant developments that have taken place in mathematics in recent times. But to include every advance that could be called significant would require several volumes, not just one, so I had to be selective—very selective. First, I restricted myself to developments that have taken place since 1960. Because the book is intended for the general reader, I included only topics that have merited attention in the world's press and that can be explained fairly simply. And of course, my own personal tastes and preferences influenced my decisions.

For the most part, all that is required of you, the reader, is an interest in the subject that caused you to pick up this book in the first place, together with a little patience. (Understanding mathematics, even elementary mathematics, takes time.) Unavoidably, a reasonably good school mathematics education would enable you to get more from some parts of my account than otherwise would be possible, but I have tried to keep these to a minimum (and you can always skip over passages you find difficult, secure in the knowledge that it will soon get "easier"). Although for the most part, the chapters are independent of one another, they are presented in such an order that earlier ones might help in the appreciation of later ones.

The first edition of this book was written in 1986 and published in 1988. In preparing this new, American edition, I have changed the text to make it better suited to American readers and to take account of various developments that have taken place in the intervening ten years, but the choice of topics remains the same. Again, my intention was never to cover all of mathematics—an impossible task in a discipline that, in terms of published new results, has, throughout my mathematical career, doubled in size roughly every ten years. Rather, I have tried to convey something of the richness and diversity of present-day mathematics.

Other than the addition of new material updating the original account and the correction of minor errors and misprints pointed out to me by various readers of the first edition, the most significant change made in this second edition is the complete rewriting of the chapter on

Fermat's last theorem to take account of the 1994 proof of the theorem. This new treatment of Fermat's last theorem required moving the chapter to follow the discussion of topology; thus, what was chapter 8 in the first edition is now chapter 10, and chapters 9 and 10 of the first edition now appear as chapters 8 and 9, respectively. The other major change is in the chapter on knots and topology, chapter 10 of the first edition and chapter 9 of this edition. This has been renamed and expanded to take account of some dramatic developments in topology that have occurred over the past five years.

Keith Devlin
May 1998
Moraga, California

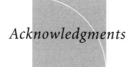

Acknowledgments

Like all mathematicians nowadays, I can call myself an expert in just one tiny area of a vast and growing landscape. Therefore, to ensure that my discussion is comprehensive, I have had to rely on others to pick up the errors that inevitably arose in my first draft. Accordingly, my thanks are due to Sir Michael Atiyah, Amanda Chetwynd, David Nelson, Carl Pomerance, Stephen Power, Hermann te Riele, Morwen Thistlethwaite, David Towers, and Robin Wilson, each of whom read all or part of the manuscript and made helpful suggestions. My thanks too to my editors at Penguin Books in the United Kingdom, who from the very start showed great enthusiasm for what must have seemed like the impossible task of trying to produce a "popular" account of humankind's most impenetrable subject. My thanks as well to Columbia University Press for its enthusiasm and support in preparing this new American edition. All failures and errors are, of course, to be laid at my door.

K. D.

Prime Numbers, Factoring, and Secret Codes

The Biggest Prime Number in the World

The biggest (known) prime number* in the world is a giant that requires 909,526 digits to write out in standard decimal format. Printing out the entire number in a book like this would require about 500 pages. Using exponential (or power) notation, however, the number has a more manageable form:

$$2^{3021377} - 1.$$

That is, you get the number by multiplying 2 by itself 3,021,376 times and then subtracting 1 from the answer.

Numbers that can be obtained by raising 2 to a power and then subtracting 1 are called *Mersenne numbers*, for reasons I will give later in this chapter. Prime numbers of this particular form are called *Mersenne primes*. Record prime numbers are almost always Mersenne numbers, because there is a particularly efficient and fairly simple way to check whether a Mersenne number is prime. (I will describe the method later.)

---

*A prime number is one that has no exact divisors other than 1 and itself. For example, 2, 3, 5, 7, and 11 are primes, but 4, 6, 8, 9, and 10 are not. I'll explain this concept in more detail later.

Exponential notation is deceptive. To try to obtain some idea of its power for representing large numbers, imagine taking an ordinary 8 × 8 chessboard and placing piles of counters 2 mm thick (a quarter is fairly close) on the squares according to the following rule: Number the squares from 1 to 64, as in figure 1. On the first square, place 2 counters; on square 2, place 4 counters; on square 3, place 8 counters; and so on, on each square, placing exactly twice as many counters as on the previous one. Thus on square n you will have a pile of 2^n counters. In particular, on the last square you will have a pile of 2^{64} counters. How high do you think this pile will be? One meter? One hundred meters? A kilometer? Surely not! Well, believe it or not, your pile of counters will stretch out beyond the Moon (a mere 400,000 kilometers away) and the Sun (150 million kilometers away) and will in fact reach almost to the

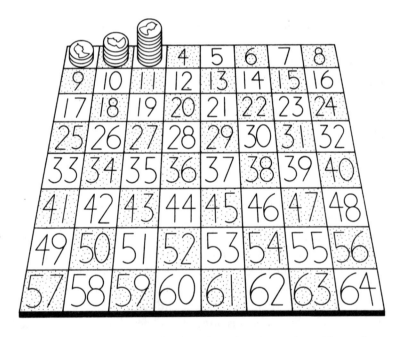

FIGURE 1   The astronomical chessboard number. By starting with two 2-mm-thick coins on the first square and forming a pile twice as high on each successive square, the pile on the sixty-fourth square will stretch almost to the nearest star, Proxima Centauri, some 4 light-years away.

nearest star, Proxima Centauri, some 4 light-years from Earth. In decimal format, the number 2^{64} is

$$18,446,744,073,709,551,616.$$

So much for 2^{64}. To obtain the number $2^{3021377}$ that appears in the record prime expression, you would need a chessboard with 3,021,377 squares—a board measuring 1739×1739 squares would do the trick.

How do you handle numbers of this size?

For a start, you use a computer. If you are working on your own, you would be advised to use one of the most powerful computers in the world. Such machines, called *supercomputers*, are capable of performing several billion* arithmetic operations a second. Even then, most of the record primes discovered in recent years have been found using supercomputer searches that have run for several months.

For example, in 1996, David Slowinski and a team of computer engineers at the Cray Research engineering and manufacturing center in Chippewa Falls, Wisconsin, used a Cray T94 supercomputer system to find the 378,632-digit record prime number

$$2^{1257787} - 1.$$

The Cray T94 was the latest in a long line of Cray computers that, among them, held the world record for finding the largest-known prime number for twelve years until that run was broken later in 1996 with the discovery of the 420,921-digit prime number

$$2^{1398269} - 1.$$

A surprising feature of this discovery was that it was made using a standard desktop personal computer, belonging to Joel Armengaud, a computer programmer working for Apsylog in Paris.

How was Armengaud's PC able to beat one of the world's most powerful supercomputers? Teamwork is the answer. Armengaud was a participant in a worldwide project known as GIMPS (Great Internet Mersenne Prime Search). Started in 1995 by George Woltman, a computer

---

*Throughout this book, "billion" means a thousand million.

programmer in Orlando, Florida, it brought together hundreds of volunteers spread around the globe in a systematic effort to discover new record primes. Each member of the team was given the computer program used to test Mersenne numbers for primality, along with a small collection of Mersenne numbers to test.

When Armengaud made his discovery, GIMPS had about 700 members, a number that had grown to more than 2000 by August 1997, when the Englishman Gordon Spence made the second GIMPS discovery, the 895,932-digit prime number

$$2^{2976221} - 1.$$

Then in January 1998, a 19-year-old California student named Roland Clarkson found the record prime given at the start of the chapter. By then, Clarkson was one of 4000 GIMPS members. All three GIMPS discoveries were independently checked by former record holder David Slowinski of Cray Research, using a Cray T90 supercomputer to repeat the calculation.

Why does anyone bother? Why do manufacturers of highly expensive supercomputers devote hours of computer time trying to find large prime numbers? Why do ordinary, seemingly sane, people all over the world set their desktop PCs whirring for weeks on end searching for these arithmetical monsters? What is so special about prime numbers anyway?

Read on, and all will be revealed.

Prime Numbers

"That action is best, which procures the greatest happiness for the greatest numbers," wrote Francis Hutcheson in 1725 (*Inquiry into the Original of Our Ideas of Beauty and Virtue*, treatise 2, section 3.8). It seems unlikely that Hutcheson was thinking of numbers in the mathematical sense of greatest-known primes and the like, but his statement nevertheless applies quite well to our never-ending fascination with those most fundamental of mathematical objects—the *natural* (or *counting*) *numbers*, 1, 2, 3, ... These abstract mathematical objects are fundamental not only to our everyday life but also to practically all mathematics—so much so that the nineteenth-century mathematician Leopold Kronecker wrote (of mathematics): "God created the natural numbers, and all the rest is the work of man."

Various properties apply to natural numbers, splitting them into two classes (those with a property and those without). For instance, one property is that of being even, which splits the natural numbers into the class of those that are even (2, 4, 6, 8, . . .) and those that are not (the odd numbers: 1, 3, 5, 7, . . .). Or there is the property of being divisible by 3. (Here, as elsewhere in this book, when we say that one number *divides* another, we mean that it does so exactly, leaving no remainder. Thus 3, 6, 9, and 12 are divisible by 3, whereas 1, 2, 4, 5, and 7 are not.) The even-odd split is a natural and important one, but the split into those numbers divisible by 3 and those not divisible by 3 is not so natural nor of great importance. Another example of a natural and important split is the property of being a *perfect square,* like $1 = 1^2$, $4 = 2^2$, $9 = 3^2$, 16, 25, 36, ... But by far the most important way of dividing up the natural numbers is into those that are prime and those that are not.

A natural number n is said to be a *prime number* if the only numbers that divide it are 1 and n itself.

Thus 2, 3, 5, 7, 11, 13, 17, and 19 all are primes, whereas 1, 4, 6, 8, 9, 10, 12, 14, 15, 16, 18, and 20 are not. (Numbers that are not prime are sometimes called *composite.* The number 1 is a special case here, and it is conventional to regard 1 as neither prime nor composite.) For instance, 7 is prime because none of the numbers 2, 3, 4, 5, and 6 divides it; 14 is not prime, since both 2 and 7 divide it.

The main reason that the prime numbers are so important was already known to the Greek mathematician Euclid (ca. 350–300 B.C.) who, in book 9 of his *Elements* (a thirteen-volume compilation of all the mathematical knowledge then available) proved what is nowadays known as the *fundamental theorem of arithmetic,* that every natural number greater than 1 is either prime or else can be expressed as a product of primes in a way that is unique except for the order in which the primes are arranged.

For instance, the number 75,900 is a product of seven *prime factors* (two being *repeated factors*):

$$75,900 = 2 \times 2 \times 3 \times 5 \times 5 \times 11 \times 23.$$

The expression on the right of the equals sign here is called the *prime factorization* of the number 75,900.

The fundamental theorem of arithmetic tells us that the prime numbers are the basic building blocks from which all natural numbers are constructed. As such, they are like the chemist's elements or the physicist's fundamental particles. Knowledge of the prime factorization of any number gives the mathematician almost complete information about that number, as is dramatically illustrated later on in this chapter (see the section on secret codes). But for now, what about the prime numbers themselves?

The most basic question you can ask about prime numbers is how common they are. Is there, for instance, a biggest prime number, or do the primes go on forever, getting larger and larger? At first glance, they seem to be very common. Of the first ten numbers beyond 1 (i.e., 2 to 11 inclusive), five are prime, namely, 2, 3, 5, 7, and 11, which is exactly half the collection. Of the next ten numbers, 12 to 21, three are prime (13, 17, and 19), a proportion of 0.3. Between 22 and 31, the proportion of primes is again 0.3, but for the next two groups of ten numbers, the proportion falls to 0.2. So the primes seem to "thin out" the farther you go along the sequence of natural numbers. Table 1 shows how the number of primes less than n [denoted by $\pi(n)$] varies with n for selected values of n and gives the "density" figure $\pi(n)/n$ in each case.

So, the primes become less common the higher up you go in the number sequence. But do they eventually peter out altogether? The answer is no. This was also demonstrated by Euclid, using an argument that to this day remains a superb model of elegant mathematical reasoning. To begin, imagine the prime numbers listed in order of magnitude:

$$p_1, p_2, p_3, \cdots .$$

TABLE I   The distribution of primes, showing the number of primes $\pi(n)$ smaller than n for various values of n.

n	$\pi(n)$	$\pi(n)/n$
1000	168	1.168
10000	1229	0.123
100000	9592	0.096
1000000	78498	0.078

So $p_1 = 2$, $p_2 = 3$, $p_3 = 5$, and so on. The aim is to show that this list must continue forever. To put it another way, we must demonstrate that if we are at any stage n in the list, having enumerated $p_1, p_2, \ldots, p_n$, there must be another prime in the list beyond p_n. The trick is to look at the number

$$N = p_1 p_2 p_3 \cdots p_n + 1$$

obtained by multiplying together all the primes p_1, p_2, p_3, and so on up to p_n and then adding 1 to the result. Obviously, N is bigger than p_n, so if N happens to be prime, we know that there is a prime beyond p_n, which is what we are trying to prove. If N is not prime, it must be divisible by a prime, call it p. But if you try to divide N by any of the primes $p_1, p_2, \ldots, p_n$, you will have a remainder of 1 (the same 1 that was added when we obtained N in the first place). So our p must be a different prime (necessarily greater than p_n), and again we have proved what was required. In any event, there is a prime bigger than p_n, and we can conclude that the list of primes continues forever.

Notice that we have no idea whether or not the number N is prime. If you try a few examples, you will discover that numbers of this form often are prime. For instance,

$$N_1 = 2 + 1 = 3,$$
$$N_2 = 2 \times 3 + 1 = 7,$$
$$N_3 = 2 \times 3 \times 5 + 1 = 31,$$
$$N_4 = 2 \times 3 \times 5 \times 7 + 1 = 211,$$
$$N_5 = 2 \times 3 \times 5 \times 7 \times 11 + 1 = 2311,$$

all are primes. But the next three are not:

$$N_6 = 2 \times 3 \times 5 \times 7 \times 11 \times 13 + 1$$
$$= 30,031 = 59 \times 509,$$
$$N_7 = 19 \times 97 \times 277,$$
$$N_8 = 347 \times 27,953.$$

In fact, no one knows whether infinitely many numbers of the form

$$N_n = p_1 p_2 \cdots p_n + 1$$

are prime or indeed whether infinitely many of these numbers are composite (though at least one of these two possibilities must be true, of course). This is just one of dozens of easily stated questions about prime numbers whose answer is not known.

One of the most famous unanswered questions about prime numbers is *Goldbach's conjecture*. In a letter to Leonhard Euler written in 1742, Christian Goldbach conjectured that every even number greater than 2 is a sum of two primes. For instance,

$$4 = 2 + 2,$$
$$6 = 3 + 3,$$
$$8 = 3 + 5,$$
$$10 = 5 + 5,$$
$$12 = 5 + 7.$$

Computer searches have verified Goldbach's conjecture for all even numbers up to 4×10^{11}, but to this day the conjecture has not been settled properly one way or the other.

Primality Testing

Although most of the classical problems concerning prime numbers have remained unsolved, the last few decades have seen tremendous developments in methods by which numbers may be tested to see whether or not they are prime. Methods for testing primality? you ask. But surely it is obvious how to go about it? And indeed, there is a perfectly natural, straightforward way of finding out whether or not a number is prime. Given a number, n say, you first see whether 2 divides it. If it does, then n is not prime, and that is the end of the matter. Then you try 3. If 3 divides n, then n is not prime, and again you are finished. Then try to divide n by 5. (You can skip past 4. Since 2 does not divide *n* if you have got this far, 4 cannot divide *n* either.) If 5 fails to divide *n*, you try 7. (Again, you can skip past 6, since 2 and 3 do not divide *n*.) And so on. If you get as far as $\sqrt{n}$ without finding a number that divides *n*, you will know that *n* must be prime (because if *n* were not prime, it would be a product of two numbers *u* and *v* between 1 and *n*, and either *u* or *v* will be no greater than $\sqrt{n}$).

This process is known as *trial division*. Although it works well for moderately small numbers, it becomes unwieldy if the numbers are too large. To see just how impractical it becomes, suppose you had to write a highly efficient program to run trial division on the fastest computer available. For a number of 10 digits, the program would appear to run instantaneously—the answer would appear immediately. For a 20-digit number, it would have a bit of a struggle and would take an hour or so. For a 50-digit number, it would require a staggering billion years. A 100-digit number would require this many years:

$$1,00,000,000,000,000,000,000,000,000,000,000,000$$

(there are thirty-five zeros here). This is not just a trivial calculation of a very large number. As I will explain later in this chapter, primes with between 50 and 100 digits are required for one of the most secure systems of secret coding in use today.

Just how do you decide whether a 100-digit number is prime? The best method available at the moment is a highly sophisticated technique developed around 1980 by the mathematicians Leonard Adleman, Carl Pomerance, Robert Rumely, Henri Cohen, and Hendrik Lenstra and often referred to by their initials as the APRCL test. When implemented on a very fast computer, the running time for the APRCL test is, for a 20-digit number, less than 10 seconds; for a 50-digit number, less than 15 seconds; and for a 100-digit number, less than 40 seconds. The computer can even handle a 1000-digit number if you give it a few days to work on the problem.

How does the test work? It depends on a considerable amount of highly sophisticated mathematics—mathematics way beyond a typical undergraduate degree course—so it is not possible to give a complete answer here. But it is not hard to explain the central idea behind the method, which is a simple (though very clever) piece of mathematics credited to the great French mathematician Pierre de Fermat (1601–1665).

Although only an "amateur" mathematician (he was a jurist by profession), Fermat produced some of mathematics' most clever results, even to this day. One of his observations was that if p is a prime number, then for any number a less than p, the number $a^{p-1} - 1$ will be divisible by p. For instance, take $p = 7$ and $a = 2$. Then

$$a^{p-1} - 1 = 2^6 - 1 = 64 - 1 = 63,$$

and indeed 63 is divisible by 7. Try it yourself for any values of p (prime) and a (less than p). The result is always the same.

Here, then, is a possible way of testing whether or not a number n is prime. Compute the number $2^{n-1} - 1$ and see whether n divides it. If it does not, n cannot be prime (because if n *were* prime, then according to Fermat's observation, you would have divisibility of $2^{n-1} - 1$ by n). But what can you conclude if you find that n does divide $2^{n-1} - 1$? Not, unfortunately, that n must be prime, though this is likely to be the case. The trouble is that although Fermat's result tells us that n divides $2^{n-1} - 1$ whenever n is prime, it does not say that there are no composite numbers with the same property. (It is like saying that all motor cars have wheels. This does not prevent other things from having wheels—bicycles, for instance.) And in fact, some nonprimes do have the Fermat property. The smallest one is 341, which is not prime, since it is the product of 11 and 31. But if you were to check (on a computer), you would find that 341 does divide $2^{340} - 1$. (We shall see in a moment that there is no need to calculate 2^{340} in making this check.) Composite numbers that behave like primes in regard to the Fermat property are called *pseudoprimes*. So when you test for primality using the Fermat result, if you discover that n does divide $2^{n-1} - 1$, all you can conclude is that either n is prime or else it is pseudoprime. (In this case, the odds are heavily in favor of n's actually being prime. For although there are in fact an infinity of pseudoprimes, they occur much less frequently than do the real primes. For instance, there are only two such numbers smaller than 1000, and only 245 below 1 million.)

Incidentally, it makes little difference if instead of 2 you use some other number, say 3 or 5, in testing the Fermat property. Whichever number you use, pseudoprimes will prevent you from obtaining an absolute answer to your primality problem.

When using the preceding test, it is not necessary to calculate the number 2^{n-1}, a number that we have already observed to be very large for even modest values of n. All you need to do is find out whether or not n divides $2^{n-1} - 1$. This means that you may ignore multiples of n at any stage of the calculation. To put it another way, what must be calculated is the remainder that would be left if $2^{n-1} - 1$ were in fact divided by n. The aim is to see whether or not this remainder is zero, but since multiples of n will not affect the remainder, they may be ignored. Mathematicians (and computer programmers) have a standard way of denoting remainders: the remainder left when A is divided by B is written as

$$A \bmod B.$$

Thus, for example, 5 mod 2 is 1, 7 mod 4 is 3, and 8 mod 4 is 0.

As an example of the Fermat test, let us apply it to test the number 61 for primality. We need to calculate the number

$$(2^{60} - 1) \bmod 61.$$

If this is not zero, 61 is not a prime. If it is zero, 61 is either a prime or a pseudoprime (and in fact is a genuine prime, as we know already). We shall try to avoid calculating the large number 2^{60}. We start with the observation that $2^6 = 64$, and hence $2^6 \bmod 61 = 3$. Then, since $2^{30} = (2^6)^5$, we get

$$2^{30} \bmod 61 = (2^6 \bmod 61)^5 \bmod 61 = 3^5 \bmod 61$$
$$= 243 \bmod 61 = 60.$$

So,

$$2^{60} \bmod 61 = (2^{30})^2 \bmod 61 = (2^{30} \bmod 61)^2 \bmod 61$$
$$= 60^2 \bmod 61 = 3600 \bmod 61 = 1.$$

Thus,

$$(2^{60} - 1) \bmod 61 = 0.$$

Since the final answer here is 0, the conclusion is that 61 is either prime or pseudoprime, as anticipated.

At this point you may wish to try a calculation yourself. So, verify that

$$2^{10} \bmod 341 = 1,$$

then use this fact to show that

$$2^{340} \bmod 341 = 1.$$

This result tells you that the number 341 is either prime or pseudo-prime. (In this case, as mentioned earlier, 341 is in fact a pseudoprime.)

The APRCL test works by altering the Fermat test so that it cannot be "fooled" by a pseudoprime. It is this alteration that requires so much deep mathematics. (If you really want to see for yourself, the place to look is the article "Primality Testing and Jacobi Sums," by Cohen and Lenstra, in the mathematical research journal *Mathematics of Computation* 42 (1984):297–330.)

Mersenne Primes

The APRCL test is the fastest general-purpose primality test currently available. The phrase *general purpose* here means that it will work on any given number n. But for numbers having special structures, alternative methods are often much faster, the process speeded up by exploiting the special structure of the number. The most spectacular example of this works for numbers of the form $2^n - 1$. Such numbers are nowadays called *Mersenne numbers* after a seventeenth-century French monk, Marin Mersenne.

In the preface of his book *Cogitata Physica-Mathematica,* published in 1644, Mersenne stated that the number

$$M_n = 2^n - 1$$

is prime for $n = 2, 3, 5, 7, 13, 17, 19, 31, 67, 127$, and 257 and composite for all other n less than 257. How did he do it? No one knows. At any rate, he was astonishingly close to the truth. Only in 1947, when desktop calculators became available, was it finally possible to check his claim. Mersenne had made just five mistakes: M_{67} and M_{257} are not prime, and M_{61}, M_{89}, and M_{107} are prime.

Mersenne numbers provide an excellent method of obtaining very large prime numbers. The rapid growth of the function 2^n as n gets larger guarantees that the Mersenne numbers M_n soon become extremely large, so the idea is to look for values of n for which M_n is prime. Such primes are called *Mersenne primes*. A bit of elementary algebra indicates that M_n is not prime unless n itself is prime, so it is necessary to look only at prime values of n. But even most primes n give rise to a composite Mersenne number M_n, so the search for suitable values of n is not easy—though this is not at all apparent from the first few cases, since

$$M_2 = 2^2 - 1 = 3,$$
$$M_3 = 2^3 - 1 = 7,$$
$$M_5 = 2^5 - 1 = 31,$$
$$M_7 = 2^7 - 1 = 127,$$

all are prime. But then the pattern breaks, with

$$M_{11} = 2047 = 23 \times 89.$$

Then follow three more prime values:

$$M_{13} = 8191, \quad M_{17} = 131{,}071, \quad M_{19} = 524{,}287.$$

After that, Mersenne primes become harder to find. The next five values of n for which M_n is prime are $31, 61, 89, 107$, and 127.

When they see these values for the first time, most people jump to the conclusion that if p is itself a Mersenne prime, M_p also must be prime. It certainly works at first: 3 is a Mersenne prime and so is M_3; 7 is a Mersenne prime and so is M_7; 31 is a Mersenne prime and so is M_{31}; likewise for 127 and M_{127}. But there the pattern stops. Although 8191 is a Mersenne prime (being M_{13}), M_{8191} (which has 2466 digits) is composite. This was discovered in 1953 using an early computer (see the section on perfect numbers later in this chapter).

In fact, to date, there are only thirty-seven known Mersenne primes. The twelve values of n just listed for which M_n is prime were all known by the early years of this century. The next five ($n = 521, 607, 1279, 2203$, and 2281) were found in 1952 by Raphael Robinson using the SWAC computer. The value $n = 3217$ was discovered in 1957 by Hans Riesel using a BESK computer. In 1961, Alexander Hurwitz used an IBM 7090 computer to obtain the values $n = 4253$ and 4423, and in 1963 Donald Gillies and the ILLIAC-II found $n = 9689, 9941$, and 11213. Bryant Tuckerman's IBM 360-91 bagged $n = 19937$ in 1971. With the next discovery, in 1978, record prime numbers became frontpage news with the announcement that after three years' work involving 350 hours of computer time on the CYBER 174 at California State University at Hayward, two 18-year-old high school students, Laura Nickel and Curt Noll, had found the 6533-digit Mersenne prime M_{21701}. One year later, Noll bettered the record with the 6987-digit prime M_{23209}. Later the same year, the record fell again, this time to David Slowinski, a young programmer working for Cray Research in Chippewa Falls, Wisconsin. Using the immensely powerful Cray-1 computer, he found the 13,395-digit prime M_{44497}. In 1982 the same man-machine combination showed that M_{86243} (a 25,962-digit number) is prime. Then, moving on to the even more powerful Cray-XMP computer, Slowinski went even higher in 1983 with the 39,751-digit prime M_{132049}. In September 1985, in Houston, Texas, a Cray-XMP owned by Chevron Geosciences found the 65,050-digit prime number M_{216091}. (Because Chevron was running Slowinski's Prime Finder program, the credit for the discovery really goes to him.)

Three years later, in early 1988, the same computer found another Mersenne prime, M_{110503}. The number is considerably smaller than the record holder discovered earlier, and its discovery raised doubts about the statistical predictors that had been used to target likely ranges of exponents for which Mersenne numbers might be prime.

Then in 1992, the record for the largest known prime crossed the Atlantic to England, with the discovery of the 227,932-digit prime number M_{756839}. The computer that was used to perform the massive calculation necessary to search for and identify the new prime was a

Cray-2 owned by Cray Research, again running Slowinski's Prime Finder program.

Early in 1994, the record moved back to the United States when Slowinski and fellow Cray computer scientist Paul Gage found the new prime M_{859433}, a number having 258,716 digits. The computer they used, a Cray C916 at the Cray Research facility, took about thirty minutes to carry out the computation verifying that this particular number was prime.

Two years later, in 1996, Slowinski and Gage found another Mersenne prime, breaking their previous record with the 378,632-digit $M_{1257787}$. But their glory was short-lived. Within a few months, Armengaud made his dramatic announcement.

The new champion, $M_{1398269}$, was the thirty-fifth Mersenne prime to be discovered. Armengaud found it by using the same test for primes used by Slowinski and Gage. But whereas the two Cray researchers used a single supercomputer, the Frenchman used an implementation of the test developed by GIMPS originator Woltman, which divided up the search among hundreds of volunteers via the Internet. In the final step, it took Armengaud's 90-megahertz Pentium personal computer 88 hours to prove his number to be a prime. The find was double-checked by others on two different computers.

In August 1997, Gordon Spence took the record back to Britain with his discovery that the 895,932-digit Mersenne number $M_{2976221}$ was prime.

Finally (for now), in January 1998, Roland Clarkson brought the record back to the United States with his GIMPS discovery of the 909,526-digit Mersenne prime $M_{3021377}$.

Why do they do it? In the case of the thousands of amateurs who signed up for Woltman's GIMPS program, the answer is obvious: for the sheer fun of being in on the search. Spence put it this way: "I saw it as my chance to make my (tiny) mark in the history books," adding the teaser, "Wouldn't you like to be the person who discovers the first million-digit prime number?"

Okay, so having fun explains the interest of amateurs, even if not everyone would see the attraction. But why does a large supercomputer manufacturer like Cray Research invest so much money in what, from its perspective, is surely little more than a game?

The answer is that the computation required to search for large Mersenne primes is a heavy one, stretching over days or weeks, and so it provides an excellent way to test the efficiency and accuracy of a new computer system. The question that interests Cray is, Does their latest computer perform the way it is supposed to?

Computer chip manufacturer Intel also uses a Mersenne prime–hunting program to test every Pentium chip before it ships it. If it is possible to have a little fun at the same time as they put the computer through its paces, then why not?

Is that the end of the story? Almost certainly not. The GIMPS project continues to attract new members, and the supercomputer manufacturers are still in the hunt. Mathematicians speculate that the number of Mersenne primes is infinite, so there should always be a larger one waiting to be discovered—although no one has been able to prove conclusively that there are infinitely many Mersenne primes. In fact, all we know for certain is that there are at least thirty-seven of them, that is, the ones actually identified so far.

With the history under our belts now, it's time to take a look at the methods used to decide whether large numbers (including large Mersenne numbers) are prime.

The method used to check Mersenne numbers for primality is very simple (although the mathematics behind it is not). It is known as the *Lucas-Lehmer test,* after Edouard Lucas (who discovered the basic idea in 1876) and Derrick Lehmer (who refined the method in 1930). To test whether the Mersenne number M_n is prime (assuming n is already known to be prime), calculate the numbers $U(0), U(1), \ldots, U(n-2)$ according to the following rules:

$$U(0) = 4,$$

$$U(k+1) = [U(k)^2 - 2] \bmod M_n.$$

If at the end you find that $U(n-2) = 0$, then M_n is prime. If $U(n-2) \neq 0$, then M_n is not prime.

For example, suppose we wanted to use the Lucas-Lehmer test to see whether M_5 is prime. (Since $M_5 = 2^5 - 1 = 31$, we know already in this simple case that it is a prime, but it will illustrate the method.) Then we make the calculation

$$U(0) = 4,$$
$$U(1) = (4^2 - 2) \bmod 31 = 14 \bmod 31 = 14,$$
$$U(2) = (14^2 - 2) \bmod 31 = 194 \bmod 31 = 8,$$
$$U(3) = (8^2 - 2) \bmod 31 = 62 \bmod 31 = 0.$$

Since $U(3) = 0$, M_5 must be prime.

You might want to try this out for yourself using the two numbers $M_7 = 127$ (which is prime) and $M_{11} = 2047$ (which is not prime—see earlier).

Factoring

At the October 1903 meeting of the prestigious American Mathematical Society, the mathematician Frederick Nelson Cole was listed in the program as presenting a paper with the rather unassuming title "On the Factorization of Large Numbers." When called on to speak, Cole walked up to the blackboard and, without saying a word, performed the calculation of 2 raised to the power of 67, after which he subtracted 1 from the result. Still saying nothing, he moved to a clean part of the board and multiplied together the two numbers

193,707,721 and 761,838,257,287.

The answers to the two calculations were the same. Cole returned to his seat still having uttered not one word, and for the first and only time on record, the entire audience at an American Mathematical Society meeting rose and gave the "speaker" a standing ovation.

What Cole had done (and apparently it took him twenty years of Sunday afternoons) was to find the prime factors of the Mersenne number M_{67}. It had been known since 1876 that M_{67} was composite. But this had been discovered (by Edouard Lucas himself) using the Lucas (now Lucas-Lehmer) test, which, although it answers the question of whether a given Mersenne number is prime or composite, gives no information about the factors of any number found to be composite. (The same is true of the APRCL test, as can be appreciated from the outline of this method given earlier, and indeed for several other fast primality-testing methods currently available.)

How do you go about finding the factors of a number that you know is composite? Trial-and-error is clearly out of the question for the same reason that it is not practicable as a test for primality. But in practice there is an element of trial division in all current implementations of primality tests and factoring methods. Because it can be done quickly, it makes sense to start by using trial division with, say, the first million prime numbers. If you find a divisor, you will have solved both the primality and the factorization problems. If you do not, at least you will know that the number is either prime or else, if composite, has only large prime factors. This last fact is used in a simple factoring method credited to Fermat and described next.

Suppose that $n = uv$, where both u and v are large, odd numbers, say $u \le v$. (Because we are assuming that n has only large prime factors, this is the situation that will face us when we know that n is composite and want to find the factors.) Let

$$x = \tfrac{1}{2}(u + v), \quad y = \tfrac{1}{2}(u - v).$$

Then $0 \le y < x \le n$, and $u = x + y$, $v = x - y$, so

$$n = (x + y)(x - y) = x^2 - y^2,$$

which can be rewritten as

$$y^2 = x^2 - n. \tag{1}$$

Conversely, if x and y satisfy equation (1), n has the factorization

$$n = (x + y)(x - y). \tag{2}$$

Hence, factoring n into a product of two numbers is equivalent to finding numbers x and y that satisfy equation (1), in which case the resulting factorization is given by equation (2). (Notice that this does not necessarily yield the *prime* factorization of n. But once a number has been split into two factors, they in turn may be factored, a task that is invariably much easier, since the smaller a number is, the easier it is to factor.)

To find x and y as in equation (1), start with the smallest number k such that $k \geq \sqrt{n}$, and then try each of the values $x = k$, $x = k + 1$, $x = k + 2, \ldots$ in turn, checking each time to see whether $x^2 - n$ is a perfect square. Once such an x is found, the factorization is effectively completed. If n has two factors of approximately the same size (hence close to $\sqrt{n}$ where the method starts), you should find a solution fairly quickly. If you want to try the method yourself at this stage, the numbers 10,379 and 93,343 provide good examples.

There are various ways of speeding up this process. For instance, if you are doing it by hand, there is no need to evaluate the square root of $x^2 - n$ in every case to see whether it turns out to be a whole number. Since no perfect square ends in any of the digits 2, 3, 7, or 8, whenever $x^2 - n$ is found to end with such a digit, you can immediately ignore that value of x.

Fermat himself used this method to obtain the factorization

$$2{,}027{,}651{,}281 = 44{,}021 \times 46{,}061.$$

A useful variant of Fermat's method looks not for numbers x, y such that

$$n = x^2 - y^2$$

but for numbers x, y such that

$$n \text{ divides } x^2 - y^2$$

Given such numbers, it is likely that $x + y$ and $x - y$ each contain a factor of n. For example, 21 divides $10^2 - 4^2$, and the two factors $10 + 4 = 14 = 2 \times 7$ and $10 - 4 = 6 = 2 \times 3$ give the factorization $21 = 7 \times 3$.

In the 1920s, Maurice Kraitchik suggested that suitable values for x and y could be obtained by patching together smaller numbers. For example, if we are trying to factorize $n = 111$, a search will reveal that

$$11^2 \bmod 111 = 121 \bmod 111 = 10 = 2 \times 5$$

$$14^2 \bmod 111 = 196 \bmod 111 = 85 = 5 \times 17$$

$$16^2 \bmod 111 = 256 \bmod 111 = 34 = 2 \times 17$$

which implies that 111 divides $(11 \times 14 \times 16)^2 - (2 \times 5 \times 17)^2$, which equals 2634×2294. The factors of 111 can then be found by a simple method for finding common factors known to the ancient Greeks, called the *Euclidean algorithm* (described on pages 146–147). Applying this method to the pair 111 and 2294 gives

$$2294 \bmod 111 = 74$$

$$111 \bmod 74 = 37$$

$$74 \bmod 37 = 0$$

and hence 37 divides both 111 and 2294. Similarly, 3 is a common factor of 111 and 2634. Thus $111 = 3 \times 37$.

Of course, just as the original Fermat method requires that $x^2 - n$ be a perfect square, the Kraitchik approach requires that the values of x emerging from the search ($11 \times 14 \times 16$ in the preceding example) lead to a nontrivial factorization of n when the Euclidean algorithm is applied.

Computer implementations use rather sophisticated methods for "instantly eliminating" impossible values of x (a process known as *sieving*). In 1974, mathematicians at the University of California at Berkeley built a specially designed electronic device for sieving numbers, the SRS-181, which could process 20 million numbers a second.

Fermat Numbers

The *n*th *Fermat number* is obtained by raising 2 to the power *n*, raising 2 by that number, and adding 1 to the result; that is,

$$F_n = 2^{2^n} + 1.$$

Thus $F_0 = 3$, $F_1 = 5$, $F_2 = 17$, $F_3 = 257$, and (already the rapid growth of these numbers due to the repeated application of the exponential function is becoming apparent) $F_4 = 2^{16} + 1 = 65{,}537$.

Interest in these numbers arose because of a claim made by Fermat in a letter written to Mersenne in 1640. Having noted that each of the numbers F_0 to F_4 is prime, Fermat wrote: "I have found that numbers of the form $2^{2^n} + 1$ are always prime numbers and have long since signified to analysts the truth of this theorem." This remark should serve as a warning to all who come to a conclusion on the basis of a small amount of information. For all his great abilities with numbers, Fermat was wrong in his claim, as was first shown conclusively by the great Swiss mathematician Leonhard Euler in 1732: $F_5 = 4,294,967,297$ is not prime. Although Euler obtained this result by trial division, the extra irony here is that a straightforward calculation using Fermat's own test demonstrates the nonprimality of F_5. The test is, remember, that if *p is prime*, then $3^{p-1} \mod p = 1$, but for $p = F_5$ you get $3^{p-1} \mod p = 3,029,026,160$, so F_5 cannot be a prime.

Subsequent work has shown just how wrong Fermat was. It is now known that F_n is composite for all values of n from 5 to 23, as well as for various other values, and the current guess is that F_n is composite for *all* values of n greater than 4.

Fermat numbers provide another example of numbers whose special form makes it possible to test for their primality in an efficient manner—one popular method being *Proth's theorem:* the Fermat number F_n is prime if, and only if,

$$3^{(F_n-1)/2} \mod F_n = -1$$

This result offers a very efficient test for the primality of a Fermat number. (It is, as you may have guessed, closely related to the Fermat test discussed earlier.) This is the method used to show that F_{22} is composite. (It had been known for some time that the Fermat number F_{23} had a small prime factor, but the proof that F_{22} is composite is relatively recent.) The proof, using Proth's theorem, was carried out in 1993 by Richard Crandall and Josh Doenias at NeXT Computer, in Redwood City, California; Chris Norrie at Amdahl Corporation in San Jose, California; and Jeff Young at Cray Research, in Eagan, Minnesota. The computation used an Amdahl supercomputer and several powerful workstations and took several months to complete. In the end, it involved more than 10^{16} basic arithmetic operations. The team estimated that a similar computation for the next Fermat number whose status is unknown, F_{24}, would take about ten years.

However, our interest here lies not in the testing for the primality of Fermat numbers, but in the factorization of the ones known to be composite, for it is in this area that some significant developments have been made in recent years, developments with applications outside the realm of mathematics. (See the section on secret codes later in this chapter.)

The Fermat number F_5 was, as we have already noted, proved to be composite by Euler, who also calculated a prime factor, 641. In 1880, a man by the name of F. Landry showed that F_6 was composite, finding the prime factor 274,177. With F_7, however, the story was somewhat different. It was proved to be composite by J. C. Morehead and A. E. Western in 1905, but it was not until 1971, when John Brillhart and Michael Morrison (armed with an IBM 360-91 computer) found the factorization

$$F_7 = 2^{128} + 1$$

$$= 340{,}282{,}366{,}920{,}938{,}463{,}463{,}374{,}607{,}431{,}768{,}211{,}457$$

$$= 59{,}649{,}589{,}127{,}497{,}217 \times 5{,}704{,}689{,}200{,}685{,}129{,}054{,}721.$$

To do this, they used a method suggested much earlier by Derrick Lehmer and Don Powers involving continued fractions. The computation took about an hour and a half.

The same two men who showed that F_7 was composite in 1905, Morehead and Western, also found in 1909 that F_8 was composite. It was only in 1981 that Richard Brent and John Pollard found the factorization. The computation took two hours on a UNIVAC 1100/42 computer. The method, devised by Pollard himself and generally referred to as the *rho method,* was at the time unusual in that unlike most methods in mathematics, it did not guarantee to produce a result. All that could be concluded from the background mathematics was that if a certain computation was performed, then it was highly likely that a factorization of the number would result within a reasonable length of time, but there was a small possibility that this would not happen. The method therefore is not like trial division, in which the chance of getting an answer within a billion years is small. There is still an element of chance, but what was clever about Pollard's method was that the odds were heavily stacked in the favor of the would-be factorer. In recent years, a number of so-called Monte Carlo methods, like Pollard's factorization technique, have been used, which trade the certainty of a result for a high probability of a result in much less time.

The two prime factors of F_8 (which has 78 digits) are

$$1,238,926,361,552,897$$

and

$$93,461,639,715,357,977,769,163,558,199,606,896,584,$$
$$051,237,541,638,188,580,280,321.$$

An intriguing development led to the complete factorization of the 155-digit F_9 in 1990. The method used was Pollard's *number field sieve.* An important feature of this method is that it replaces the task of factoring the given large number with a large set of much smaller factorization problems. These smaller—but still challenging—factorization problems were farmed out to different computers spread all over the world, much like the GIMPS search for record prime numbers, but with the participants in this game being almost exclusively professional mathematicians.

The work was organized by two mathematicians, Mark Manasse of the Digital Equipment Corporation in Palo Alto, California, and Arjen Lenstra of the Bellcore Research Laboratory in New Jersey. After performing the initial computation that broke the problem into many smaller factorization tasks, they used the Internet to parcel out the individual smaller problems to colleagues having suitably powerful desktop workstations. Each colleague set his or her personal computer to work on a number of the smaller factorizations, using a factorization program supplied by the organizers. The different computers communicated over the Internet.

After two months' work, enough team members had obtained a factorization for Manasse and Lenstra to try to piece together all the results thus far into the sought-after factorization of F_9. This task, itself a huge computation, was performed on a massively parallel supercomputer known as a *connection machine,* located at the Supercomputer-Computational Research Institute in Florida. This final computation took three hours to complete but in the end was successful: F_9 was split into three factors, a 7-digit prime factor, which had been known for many years, and two more primes, one having 49 digits and the other 99 digits.

Success was the result of a novel combination of ingenious mathematics, large powerful mainframe computers, a large number of powerful desktop computers spread around the world, and the Internet, which enabled all these different computers to communicate with one another.

The next two Fermat numbers, F_{10} and F_{11}, have also been completely factored, by Richard Brent using another relatively new method, the *elliptic curve method*. Brent's results provide yet another illustration of the incredible power resulting from a marriage of sophisticated mathematics and raw computing capacity. Had he had access to fast computers, who knows what the eighteenth-century German mathematician Karl Friedrich Gauss would have achieved in regard to Fermat numbers.

Gauss certainly produced what must surely be the most astonishing result on Fermat numbers, linking them with a classical problem of Greek geometry. This result, like many of Gauss's discoveries, merits a special introduction to one of the most amazing mathematical minds the world has ever known.

An Amazing Mathematical Mind

Karl Friedrich Gauss was born in Brunswick, Germany, in 1777. His father, a bricklayer, hoped that his son would be able both to assist him in his work and to keep the accounts. The latter task was one to which the young Gauss seemed particularly well suited when, as a 3-year-old child, he was able to correct his father's payroll calculations. Fortunately for the future of mathematics (to say nothing of physics and astronomy), the reigning duke soon heard about the young child prodigy and arranged for his formal education. At 15 years of age, having progressed far beyond the abilities of his schoolteachers, Gauss went to Caroline College. Within three years, even the professors there had to admit he had left them way behind as well.

It was while he was a college student, in 1796, that Gauss made his remarkable observation concerning Greek geometry and Fermat numbers. The result appeared in the seventh and last section of his mammoth work *Disquisitiones Arithmeticae,* a book—still in print to this day—that appeared in 1801 when Gauss was still only 24 and that forms the basis of much of present-day number theory (the branch of mathematics that deals with properties of natural numbers, of which the material in this chapter is but a small part).

One of the favorite problems of the ancient Greek mathematicians was the construction of plane figures (circles, triangles, parallelograms, and so on) using only a ruler (unmarked and thus suitable only for drawing straight lines) and a compass (used only to draw arcs of circles, not to transfer a length by moving the instrument across the page). By using often considerable ingenuity, it is possible to construct many geometrical figures using just these two rudimentary tools. (Until the mid-1960s, such constructions formed a significant part of the mathematical schooling of pupils around the world.) The Greeks themselves knew how to construct regular n-sided polygons for $n = 3, 4, 5, 6, 8, 10, 12, 15$, and 16. (A polygon is *regular* if all its sides have the same length and all its internal angles are equal.)

The 19-year-old Gauss proved that you can construct a regular polygon with n sides using only a ruler and a compass if, and only if, either $n = 2^k$ for some number k or else $n = 2^k p_1 p_2 \ldots p_r$ (for some k) where $p_1, p_2, \ldots, p_r$ are distinct Fermat primes. In particular, for any Fermat prime p, you can construct a regular polygon with p sides. For the first Fermat prime, $F_0 = 3$, you get an equilateral triangle, which is easy to construct, and for the next one, $F_1 = 5$, you get a regular pentagon. Since $F_2 = 17$ is also a Fermat prime, Gauss's result shows that a regular 17-sided polygon may also be constructed using a ruler and compass. This was the first (and only) advance on the problem of constructing regular polygons since the time of the Greeks themselves, and Gauss was so proud of his discovery that he asked for a regular 17-sided polygon to be engraved on his tombstone. Although this request was never fulfilled, such a polygon is inscribed on the side of a monument to him erected in Brunswick, his birthplace.

Perfect Numbers

As was noticed by the Pythagoreans (the followers of the sixth-century B.C. mathematician Pythagoras), the number 6 has a rather special property. It is equal to the sum of its own divisors (other than itself):

$$6 = 1 + 2 + 3.$$

The next number after 6 with this property is 28. The only numbers that divide 28 are 1, 2, 4, 7, 14, and 28 itself, and

$$28 = 1 + 2 + 4 + 7 + 14.$$

Such numbers were named *perfect numbers* by the Pythagoreans.

In his first-century A.D. *Introductio Arithmeticae,* the Greek mathematician Nicomachus lists four known perfect numbers, the third (after 6 and 28) being 496 and the next, 8128. Two conjectures followed from this evidence, that the nth perfect number contains n digits and that the perfect numbers end alternately in 6 and 8. Both conjectures are false, however. For a start, there are no perfect numbers with five digits. Moreover, although the fifth perfect number does end in a 6 (it is 33,550,336), so does the sixth, which is 8,589,869,056. (It is true, however, that any even perfect number ends in either 6 or 8. This can be proved directly and does not depend on knowing which numbers actually are perfect.)

In book 9 of his *Elements,* Euclid proved in about 350–300 B.C. that if $2^n - 1$ is prime, then the number $2^{n-1}(2^n - 1)$ is perfect. Two thousand years later, Euler showed that every even perfect number is of this type. Thus was established the close relationship between Mersenne primes and perfect numbers, which implies that there are at present exactly 37 known even perfect numbers. But there are no known odd perfect numbers, and it is conjectured that all perfect numbers are necessarily even. Although this has not been proved, some evidence has been collected in favor of the conjecture. It is known that any odd perfect number, if one were to exist, would have to be larger than 10^{100} and have at least 11 distinct prime factors. On the other hand, if history is any guide, one should be careful about making conjectures about perfect numbers. In his 1811 book *Theory of Numbers,* Peter Barlow wrote in reference to the eighth perfect number, $2^{30}(2^{31} - 1)$, a 19-digit number discovered by Euler in 1772: "[It] is the greatest that ever will be discovered; for as they are merely curious, without being useful, it is not likely that any person will ever attempt to find one beyond it."

Although Barlow appears to have been right about perfect numbers being merely of curiosity value, he clearly underestimated the fascination of curiosities, as the first section of this chapter illustrates only too well. And there is no doubt that perfect numbers are very curious. For instance, every (even) perfect number is *triangular*, which means that it can be represented by a number of balls arranged to form an equilateral triangle (which is equivalent to being of the form $\frac{1}{2}n(n + 1)$ for some number n). Another fact: for any perfect number other than 6, if you add together all its digits, the number you get will be one more than a multiple of 9. Related to this is the result that the *digital root* of any perfect number is 1. (To obtain the digital root, you add together all the digits of the number, then all the digits of *that* number, and so on until you end up with a single-digit number.)

Again, every perfect number is a sum of consecutive odd cubes. For instance,

$$28 = 1^3 + 3^3,$$

$$496 = 1^3 + 3^3 + 5^3 + 7^3.$$

One more: if n is perfect, then the sum of the reciprocals of all the divisors of n is always equal to 2. For instance, 6 has divisors 1, 2, 3, 6, and

$$\frac{1}{1} + \frac{1}{2} + \frac{1}{3} + \frac{1}{6} = 2.$$

In fact, so much effort has been spent searching for these "curious" numbers that despite Barlow's claim about their uselessness, their computation has acquired something of the status of a benchmark for measuring computer power. An example is the Mersenne number M_{8191}, the first number to break the chain of Mersenne primes giving rise to Mersenne primes. Using the Lucas-Lehmer test to demonstrate that this 2466-digit number is not prime (and hence does not yield a perfect number) took 100 hours when first done on the ILLIAC-I computer in 1953. Over the years, the computation time has been reduced dramatically: 5.2 hours on an IBM 7090, 49 minutes on ILLIAC-II, 3.1 minutes on an IBM 360-91, 10 seconds on a Cray-1, and virtually no time at all on today's fastest supercomputers.

Secret Codes

In the autumn of 1982, at a scientific meeting in Winnipeg, Canada, two mathematicians and a computer engineer went out for a beer one evening. The two mathematicians soon got around to talking about how you can factorize large numbers and the computational problems that can arise. Hearing them, the computer engineer mentioned that the design of the particular machine he worked on would enable one of the major problems they had met with to be overcome quite easily. And so it was that a chance encounter in a bar had significant repercussions in the field of data security. For the difficulty of factoring large numbers lies at the heart of one of the most secure forms of secret code. The story of just how an apparently useless and esoteric piece of pure mathematics came to be the basis of modern security systems is one of the most interesting mathematical tales of this century and a dramatic warning to anyone who declares that a particular piece of scientific work is "of no practical use."

Incidentally, some of the worst offenders when it comes to minimizing the usefulness of their subject are mathematicians themselves. Writing in his excellent little book *A Mathematician's Apology,* the great British mathematician G. H. Hardy noted, "Real mathematics has no effects on war. No one has yet discovered any warlike purpose to be served by the theory of numbers or relativity, and it seems very unlikely that anyone will do so for many years" (chapter 28). This was written in 1940. By 1945 the world had seen the horrific disproof of Hardy's claim about warlike uses for relativity in the form of the atomic bomb. In regard to his other example, number theory, this "useless" subject now provides the security systems that are used to control (and perhaps one day even launch) the hundreds of nuclear missiles that have proliferated since that first atomic bomb was dropped on Hiroshima. So much for predicting applications (or lack of them) of mathematical discoveries in the world at large. Hardy's own subject was, incidentally, number theory itself, and some of his own work has proved to be of real use, despite his claim—made in chapter 29 of the same book—that "I have never done anything 'useful.' No discovery of mine has made, or is likely to make, directly or indirectly, for good or ill, the least difference to the amenity of the world."

But I digress. To get back to the consequences of that beer, notice that there is nothing new in the idea of secret codes. Julius Caesar used them to ensure the security of the orders he sent to his generals during the Gallic wars. Nowadays, it is not only the military who require their communications to be made secure by encryption techniques; there also are commercial and political reasons for ensuring that information does not pass into the wrong hands.

How do you go about designing an encryption system? A not entirely flippant answer would be, "with great care." Would-be cryptanalysts (i.e., the "enemy" trying to break your code) have many weapons at their disposal, in the form of both powerful computational equipment and sophisticated mathematical and statistical techniques. Certainly, the simple kinds of cipher that Caesar used would be woefully inadequate. In a *Caesar cipher*, the original message is transformed by replacing each letter of each word with some other letter according to some fixed rule, such as taking the letter three places farther along in the alphabet, so that A would be replaced with D, G with J, Y with B, and so on. Thus the word *mathematics* would become *pdwkhpdwlfv*. Without knowing the rule used, a message encrypted in this way may look on the surface to be totally indecipherable, but this is by no means the case. For one thing, there are only 25 such "shift along" ciphers, and an enemy who suspected you were using one would only need to try them all until he found the one you used. But even if you employ some other, less obvious rule for substituting letters, the resulting code will not be secure. The problem is that there are definite frequencies with which the individual letters occur in English (or in any other language), and by counting the number of occurrences of each letter in your coded text, an enemy can easily deduce just what your substitution rule is—especially when computers are used to speed up the process.

With simple substitution out of the question, what else might you try? Whatever you choose, the same dangers are present. If there is any kind of "recognizable" pattern in your coded text, a sophisticated statistical analysis can usually crack the code without much difficulty. Now the real difficulty becomes apparent. In order that your message can be successfully decoded at the other end (possibly thousands of miles away), the transformation performed on the message by your encryption scheme clearly must not destroy *all* order—the message must still remain beneath it all, and yet this hidden order must be buried deeply enough to prevent an enemy from discovering it.

All modern cipher systems use computers; they must. The enemy may be assumed to have powerful computers to analyze the message, so the system needs to be sufficiently complex to resist a computer attack. Because of the difficulty of designing and maintaining the security of cipher systems, they invariably consist of two components: an encryption procedure and a "key." The former is typically a computer program or possibly a specially designed computer. To encrypt a message, the system requires not only the message but also the chosen key, usually a secretly chosen number. The encryption program codes the message in a way that depends on the chosen key, so that only by knowing that key is it possible to decode the ciphered text produced (see figure 2). Because the security depends on the key, the same encryption program may be used by many people for a long period of time, which means that a great deal of time and effort can be put into its design. A helpful analogy is that manufacturers of safes and locks are able to stay in business by designing one type of lock that may be sold to hundreds of users, who rely on the uniqueness of their own key to provide security. (The "key" here could be of the combination variety, which immediately reveals the similarity between the two uses of the word *key* in this discussion.) Just as an enemy may know how your lock is designed and yet still be unable to break into your safe without knowing the combination, so the enemy may know what encryption system you are using without being able to

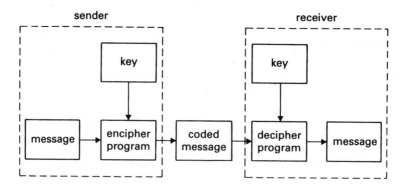

FIGURE 2   A typical cipher system. The encipher program (which may be in the form of specially made equipment or a program for a general-purpose computer) uses a secret key chosen by the user to produce the encoded text. At the receiving end, a similar system operates. Traditional systems employ the same key for both encryption and decryption, and the decipher program simply reverses the steps in the cipher program. Public-key systems utilize two different keys, and the relationship between encipherment and decipherment depends on the mathematics involved.

crack your coded messages—a task for which knowledge of your key is required.

In a typical "key system," the message sender and receiver agree beforehand on a secret key that they can use to send each other messages. As long as they keep this key secret, the system should be (if it is well designed) secure. One such system is the American-designed Data Encryption Standard (DES), which requires for its key a number whose binary representation has 56 bits (i.e., a string of 56 zeros and ones operates as the key). Why such a long key? Well, no one makes any secret of how the DES system works. All the details have been published. So in theory an enemy could crack your coded messages simply by trying all possible keys until finding the one that works. With the DES, there are 2^{56} possible keys to be tried, a number that is so large as to render the task virtually impossible. (Actually this figure is not quite high enough to provide perfect security, but with any cipher system, a trade-off must be made between security and convenience for the user. The bigger the key is, the more cumbersome the process becomes.)

Although the DES enjoys widespread use at the moment, such a system has an obvious drawback. Before it can be used, the sender and receiver must agree on the key they will use, and since they will not want to transmit that key over any communication channel, they must meet and choose the key or, at the very least, employ a trusted courier to convey the key from one to the other. Furthermore, such a system is not good for communication between individuals who have not already met. In particular, it is not suitable for use in, say, international banking or commerce, in which it is often necessary to send secure messages across the world to someone the sender has never met.

In 1975, Whitfield Diffie and Martin Hellman proposed a new type of cipher system: public-key cryptography, in which the encryption method requires not one but two keys—one for enciphering and the other for deciphering (like having a lock that requires one key to lock it and another to unlock it). Such a system is used as follows: A new user purchases the standard program (or special computer) used by all members of the communication network concerned. She then generates two keys. One of these, her deciphering key, she keeps secret. The other key, the one used for encoding messages sent to her by anyone else in the network, she publishes in a directory of the network users. To send a message to a network user, all that one does is to look up the user's public-encipherment key, encrypt the message using that key, and send it. To decode the message, it is not helpful to know (because everyone does) the enciphering key. Instead, you need the deciphering key, and only the intended receiver knows that (so even the message sender cannot decode the message once it has been enciphered).

This all sounds fine, but how can such a system be constructed? On the face of it, it seems impossible. The key (if I may use the word here) is to exploit the strengths and weaknesses of those very computers whose existence makes the encrypter's task so difficult. As was indicated earlier in this chapter, finding large primes (say on the order of 50 digits) is relatively easy, and so is the task of multiplying two such large primes together to produce a single (composite) number (of around 100 digits or more). But factoring that number into its two primes is not at all easy and, indeed, to all intents and purposes, is impossible. This is the idea behind the most common implementation of a public-key system in use today—the one designed by Ronald Rivest, Adi Shamir, and Leonard Adleman of the Massachusetts Institute of Technology and known by their initials as the RSA system. (The success of the method led to the establishment of a commercial company specializing in data security, RSA Data Security, based in Redwood City, California.)

The secret deciphering key used in the RSA method consists (essentially) of two large prime numbers chosen by the user (chosen with the aid of a computer—not chosen from any published list of primes, to which an enemy might have access). The public-encipherment key is the product of these two primes. Because there is no quick method of factoring large numbers, it is practically impossible to recover the deciphering key from the public-encipherment key. Message encryption corresponds to multiplication of two large primes (easy), and message decryption to the opposite process of factoring (hard). (This is not quite how the system works. Some mathematics is required that, although moderately sophisticated, was known to Fermat. The important point is that since deciphering a message is just the opposite of enciphering it, the same must be true of the relationship between the two keys.)

This is why the security of large international data networks nowadays relies on the inability of mathematicians to find an efficient method of factoring large numbers (while at the same time being able to produce large primes with ease). Obviously, the security of such a system depends on (among other things not mentioned here) the continued difficulty of factorization, which is where the Winnipeg beer comes in. The designers of the system originally suggested that two primes of around 50 digits each should provide sufficient security. (As always with such systems, the bigger the numbers used are, the more expensive they become to run, so a good compromise value is sought.) Until 1982, the best factoring methods developed were able to handle numbers of around 50 digits (using computers like the Cray-1). Then came the crucial barroom conversation in 1982. The occasion was a meeting at the University of Manitoba in Winnipeg. Pomerance was there to give a lecture on a new factorization method he had developed, which he had called the *quadratic sieve method.* After his talk, Pomerance got together for a drink with fellow mathematician Marvin Wunderlich and Cray computer engineer Tony Warnock. Himself a factoring expert, Wunderlich was very interested in Pomerance's new method and wanted to know more. In an earlier discussion with AT&T mathematician Andrew Odlyzko, Pomerance had wondered about implementing his new method on a Cray, making use of the particular architecture of the Cray computer. So, during the barroom conversation, Pomerance challenged Wunderlich to give his new method a try, teasing him that if such an approach were successful, it would make obsolete Wunderlich's favorite factorization technique (the *continued-fraction algorithm*). As Pomerance hoped, Warnock was hooked. Pomerance sent the details of his new method to Cray Research, where Warnock arranged for Jim Davis and Diane Holdridge to try it on the Cray supercomputer at the Sandia National Laboratories. It worked in just the way everyone had hoped. The world of computational mathematics had a new factoring method that worked.

By discovering how the particular design of the Cray-1 arithmetic units could be exploited to overcome one of the problems they were up against, Warnock gave Wunderlich and his colleague Gus Simmons just the information they needed to extend their methods to handle numbers of 60 to 70 digits. Suddenly the RSA systems looked a little less secure. Although the answer was obvious—simply use primes of 100 digits each to produce a 200-digit public key—this unexpected advance created a ripple of uncertainty in the communications security business. This feeling of "insecurity" has not been helped by other recent advances in factorization, for although 90-digit numbers appear to be the current upper limit of what is (in general) possible in factorization, the amount of sophisticated mathematics currently being brought to bear on the problem could at any time bring a real breakthrough.

That developments in factoring do indeed pose a genuine, if potential, threat to RSA codes was illustrated in April 1994, when a modern version of Kraitchik's method was used to crack a challenge problem in RSA cryptography posed in 1977.

The origin of the problem is itself of interest. In 1977, when Ronald Rivest, Adi Shamir, and Leonard Adleman proposed their public-key encryption system, it was described by mathematics writer Martin Gardner in his popular mathematics column in the August issue of *Scientific American*. Gardner presented a short message that had been encoded using the RSA scheme, with a 129-digit key resulting from the multiplication of two large primes. The message and the key were produced by researchers at MIT, who offered, through Gardner, $100 to the first person who managed to crack the code. The composite number that was the key to the code became known as RSA-129. At the time, it was thought that it would take more than 20,000 years to factor a 129-digit number of this kind, so the MIT group thought their money was safe. Two developments after 1977, however, led to the solution to the MIT challenge just seventeen years later.

The first new development was Pomerance's quadratic sieve method for factoring large numbers, mentioned already. This method builds on a systematic means of implementing the Kraitchik approach developed by Morrison and Brillhart in the 1970s.

The idea of the Kraitchik approach, remember, is to obtain the desired factorization of n by patching together a lot of small squares x^2. The question is how you find those small squares in the first place. Brillhart and Morrison's idea was to begin by specifying a collection of small primes, called a *factor base,* and then patiently trial-dividing a large number of candidate values x, keeping only those for which the prime divisors of x^2 mod n belong to the factor base. Given enough values of x, some fairly elementary algebra guarantees that some combination of them can be patched together to yield (via the Euclidean algorithm) a factorization of n. Pomerance's quadratic sieve provides an efficient method for evaluating the candidates x.

Following the introduction of the quadratic sieve, various improvements to it over the years prepared the ground for the assault on the problem of factorizing RSA—in particular, a variant of the method called the *multiple polynomial quadratic sieve*, devised independently by both Jim Davis and Peter Montgomery, which uses general quadratic polynomials $(ax + b)^2 - n$ instead of the $x^2 - n$ of Pomerance's original method. Using this new method, it is possible to run several different quadratic sieves separately, thereby allowing the computation to be divided up and run on several different machines.

It was the combination of the multiple polynomial quadratic sieve and another new development—the Internet—that facilitated the cracking of RSA-129. Taking inspiration from the factorization of F_9 carried out in 1990 by Manasse and Lenstra in 1993, Paul Leyland at Oxford University, Michael Graff at Iowa State University, and Derek Atkins at MIT put out a call on the Internet for individuals to volunteer their—and their personal computers'—time to a massive worldwide assault on RSA-129. The idea was to distribute lists of candidate small x-factors, along with the software required to process them, and then sit back and wait until enough small factorizations had been found to produce a factorization of RSA-129. This was thus a larger-scale version of the earlier Manasse-Lenstra Internet approach, with the new twist that this time anyone could join in. The programs themselves were ones developed by Manasse and Lenstra. The factor base consisted of some 524,339 primes. Some 600 volunteers, spread around the world, rose to the challenge.

Over the next eight months, results came in at the rate of around 30,000 a day. By April 1994, with more than 8 million individual results to work on, Lenstra put a MasPar ("massively parallel") supercomputer to work on the awesome task of looking for a combination of the small factorizations that would yield a factor of RSA-129. It was a mammoth computation, but in the end it was successful, with RSA-129 factored into two primes, one having 64 digits and the other 65. With it, the original MIT message was decrypted as "The magic words are squeamish ossifrage." (The ossifrage is a rare vulture having a wingspan of up to 10 feet, whose name means "bone breaker.")

Suggested Further Reading

An excellent description of number theory (though not of computational aspects) is provided by *Elementary Number Theory*, by David Burton (Allyn & Bacon, 1980). Equally good, and including computation issues, is *A Friendly Introduction to Number Theory*, by Joseph Silverman (Prentice-Hall, 1997).

At a somewhat more advanced level, try *A Course in Number Theory*, by H. E. Rose (Oxford University Press, 1994).

For a gentle coverage of cryptography, take a look at the book *Cryptology*, by Albrecht Beutelspacher (Mathematical Association of America, 1994). For a more thorough treatment, consult the book *Cryptography: Theory and Practice*, by Douglas Stinson (CRC Press, 1995).

2

Sets, Infinity, and the Undecidable

New Horizons

Sometimes the solution of a long-standing problem marks the end (or the beginning of the end) of a mathematical era or field, the culmination of years of effort. On other occasions, it may open up an entire new area of research, possibly previously undreamed of. Such was the case with the discovery in 1963, by the 29-year-old Stanford University mathematician Paul Cohen, of the solution to *Cantor's continuum problem*. Not only was the nature of the solution itself revolutionary, but also the methods Cohen developed to obtain it were new. These methods were soon found to have a wide range of possible applications, and the ensuing twenty-year period saw tremendous activity based on Cohen's breakthrough. In recognition of his work, in 1966 Paul Cohen was awarded a Fields Medal, the highest prize that can be bestowed on a mathematician, equivalent to a Nobel Prize in the other sciences.

Before 1963, a mathematician trying to determine the truth or falsity of a mathematical statement had to choose between two possibilities: either prove it true or prove it false. Experience and intuition are often the only guide to which of these is worth the greater effort, and a bad choice can lead to an immense amount of time spent trying to achieve the impossible. But it was always felt that at the end of the day, an answer would be found. What Cohen did was to destroy forever this comforting belief, by demonstrating that some mathematical statements are neither true nor false but *undecidable.* (Actually this is not quite accurate. The existence of undecidable statements was established by Kurt Gödel in 1930 but, as I will explain later, this was not thought to affect ordinary, everyday, run-of-the-mill mathematics in the way that Cohen's result did.)

To explain just what happened in 1963, we must go back to the very nature of mathematics itself and to a pioneering concept from the turn of the century.

The Axiomatic Method

Faced with determining the truth or falsity of a particular statement, the physicist, chemist, or biologist—indeed, almost any scientist—usually sets up an experiment or, at the very least, uses some reasoning that depends on experimental evidence. Each of these scientists is studying an aspect of the physical world, and consequently it is the physical world that is the final arbiter of what is and what is not. But what happens in mathematics?

At the most basic level, mathematics is much like any other physical science in that certain aspects of the world around us are singled out for detailed study. The world itself can therefore provide some information. For instance, if you want to check that the sum of the angles of a triangle is 180°, you can go out and measure the angles of lots and lots of triangles. Although this would clearly provide a verification of the kind accepted by the physicist or chemist, it is not how the mathematician proceeds in practice, nor would such a procedure be a *mathematical* verification of the assertion that the sum of the angles of *any* triangle is 180°.

The reason that a "crude," experimental approach is not adequate for determining mathematical truth lies in the nature of what mathematics is and is intended to be. Although its roots lie in the physical world, mathematics is a precise and idealized discipline. The "points," "lines," "planes," and other ideal constructs of mathematics have no exact counterpart in reality. (Chapter 4 sheds some interesting light on this point.) Rather, mathematicians adopt a totally abstract, idealized view of the world, and reason with their abstractions in a precise and rigorous fashion. A simple (but important) example should help make this clear.

A basic mathematical abstraction is that of numbers, and it is through numbers that most of us have our first encounter (usually at a very early age) with mathematical abstractions. By a process that seems almost miraculous when you come to think about it, all small children come to recognize that a collection of three apples, a collection of three uncles, a collection of three flowers, and so on all have something in common. This abstraction of "threeness" leads to the formation of the mental concept of the number three. What makes this seem miraculous is that there is not, of course, anything in the world that is the number three. Rather, it is purely an abstract concept, and it is only our familiarity with it that results in our feeling no unease (or even embarrassment) when we speak of "the number three"—or any other number. (And you will realize just how abstract the concept is if you try to explain what the number three is without using the word *three* or *threeness*. You cannot do it, yet none of us worries about that.) The same is true of all the other mathematical abstractions: although they may have their roots in the real world, the abstractions themselves are only concepts, having no existence outside our minds.

Already with the concept of numbers, we have the basis of the mathematical discipline of *number theory* discussed in the previous chapter. But how do you deal with abstractions at all, let alone in a manner that can be of use in the physical world? The only precise way is first to establish some ground rules. For numbers, this amounts to formulating *postulates* (or *axioms*) assumed to apply to all numbers and then making deductions from these initial postulates by means of rigorous, logical reasoning alone. (It also is possible to formulate postulates governing the logical reasoning process used. This is the business of the branch of mathematics known as *mathematical logic*, which is closely connected with the subject of this chapter.) For example, we know from experience that when two numbers are added together, the order in which they are added is immaterial. For example, $5 + 3$ is the same as $3 + 5$. So a reasonable axiom to include in a system for arithmetic is the statement

for any numbers m, n, $m + n = n + m$,

which is known as the *commutative law of addition*. Another example is the *associative law of addition*:

for any numbers m, n, k, $(m + n) + k = m + (n + k)$.

Again, this arises from observations of the way addition works in practice. For example, if you want to add together the three numbers 3, 5, and 10, it does not matter if you first add 3 to 5 (to get 8) and then add 10 to the result or else begin by adding 5 to 10 (to get 15) and then add 3 to that result. In either case, you get the same answer, 18.

By adopting the preceding two axioms, we are already taking a big step, a step that really amounts to nothing other than an act of faith. Although both axioms can be tested ("experimentally") by examining a large number of cases, there is not even the theoretical possibility of checking every case, since there are infinitely many of them. Can we therefore be sure that the two axioms remain true when the numbers involved are very large, perhaps having millions of digits? It seems reasonable, possibly even "obvious." But then, mathematics (and most other disciplines) is full of examples of "obvious truths" that turn out to be false. (On the basis of common experience, it does, after all, seem "obvious" that the Sun orbits the Earth.) Instead, the available evidence can only *suggest* that the two axioms are true. There is no possibility of ever being able to *prove* this conclusively; their truth must be *assumed*. That is why such assumptions are referred to as *axioms*—from the Latin word *axioma*, meaning "principle." (There is a sense in which both the preceding axioms can be proved, in that it is possible to write down "more fundamental" postulates for natural numbers and deduce the more familiar rules of arithmetic from them, but this only shifts the act of faith one step back; it does not eliminate it.)

To emphasize the point made in the preceding paragraph, I should perhaps mention that although both laws considered are accepted axioms for the arithmetic of the integers, the commutative law is false for certain systems of infinite numbers (see later in this chapter), and the associative law fails when applied to computer arithmetic. (The law fails when very large numbers are added to very small ones.)

At this stage, we now will look at the axiomatic development of a mathematical theory in a little more detail. Since we have already examined some aspects of it, the arithmetic of the integers (i.e., positive and negative whole numbers) provides an excellent example.

The Integers: An Example

The following axioms are adequate for studying the basic arithmetic (i.e., addition and multiplication) of the integers.

1. For all m, n, $m + n = n + m$ and $mn = nm$ (commutative laws of addition and multiplication).

2. For all m, n, k, $(m + n) + k = m + (n + k)$ and $(mn)k = m(nk)$ (associative laws of addition and multiplication).

3. For all m, n, k, $m(n + k) = (mn) + (mk)$ (distributive law).

4. There is a number 0 that has the property that for any number n, $n + 0 = n$ (existence of an additive identity).

5. There is a number 1 that has the property that for any number n, $n \times 1 = n$ (existence of a multiplicative identity).

6. For every number n, there is another number k such that $n + k = 0$ (existence of additive inverses).

7. For any m, n, k, if $k \neq 0$ and $km = kn$, then $m = n$ (cancellation law).

Starting with these axioms, it is possible to prove all the usual properties of integer arithmetic. For instance, there is a rule analogous to axiom 7 that applies to addition:

$$\text{if } k + m = k + n, \text{ then } m = n.$$

To prove this, start with the assumption that $k + m = k + n$. Then, by axiom 1,

$$m + k = n + k.$$

By axiom 6, let l be a number such that $k + l = 0$. Then, by adding l to both sides of the preceding equation, we get

$$(m + k) + l = (n + k) + l.$$

So, by axiom 2,

$$m + (k + l) = n + (k + l).$$

In other words, when the choice of l is taken account of,

$$m + 0 = n + 0.$$

Using axiom 4, it follows at once from this last equation that

$$m = n,$$

as required.

Again, to prove the result that $x \times 0 = 0$ for any number x, we argue as follows:

$x + 0 = x$ (by axiom 4, with $n = x$),

$\quad = x \times 1$ (by axiom 5, with $n = x$),

$\quad = x \times (1 + 0)$ (by axiom 4, with $n = 1$),

$\quad = (x \times 1) + (x \times 0)$ (by axiom 3, with $m = x$, $n = 1$, $k = 0$),

$\quad = x + (x \times 0)$ (by axiom 5, with $n = x$).

So, by the additive analogue of axiom 7 just proved (with $k = x$, $m = 0$, $n = x \times 0$),

$$0 = x \times 0.$$

At this stage, you might like to try your hand at proving each of the following standard facts about integer arithmetic. In each case, you should make sure that you use only facts already known, either because they are axioms or because you have proved them.

1. There is exactly one element 0 that satisfies the requirements of axiom 4; that is, if $\varnothing$ has the property that $n + \varnothing = n$ for every number n, then $\varnothing = 0$ (uniqueness of 0).

2. There is exactly one element 1 that satisfies the requirements of axiom 5 (uniqueness of 1).

3. For every pair m, n, there is exactly one number k such that $n + k = m$.

Notice that the last result guarantees that subtraction is always possible in the integers (since the unique number k will be $m - n$), even though this operation was not mentioned in the axioms themselves. A particular case of the result is when $m = 0$, proving the uniqueness of k in axiom 6.

Of course, if it were necessary to prove everything in mathematics in the same detail as we just did, the mathematician's task would be nearly impossible. What makes it all work is the fact that mathematical knowledge is cumulative: once something has been established, it may be used thereafter without further ado. (A very simple example of this phenomenon occurred in the preceding proof that $x \times 0 = 0$.) Consequently, it is only at the very beginning of an axiomatically based development that such detailed proofs are necessary. For the most part, mathematical reasoning is much more like a rigorous version of the "everyday logic" employed in any other science.

Consistency, Completeness, and Truth

The bulk of present-day mathematics consists of making deductions from axioms. These axioms need not relate (directly or even indirectly) to anything in the physical world. The axioms for integer arithmetic given in the previous section were obtained by examining the behavior of the operations of addition and multiplication on those integers familiar to us. (This really means the small integers, although it should be remembered that it was only in the eighteenth century that *negative* numbers became widely accepted—see chapter 3.) But once the axioms have been established, all questions about their "universal truth" become irrelevant, as do questions about just what the objects are that the axioms refer to. For instance, nowhere do the axioms just given mention what exactly a number is. In fact, many other collections of mathematical objects also turn out (as a consequence of their formal definitions) to satisfy these axioms. Because axiom systems often apply in many different situations, mathematicians frequently make up names to describe the structures satisfying a particular axiom system. For example, any mathematical structure that satisfies the axioms of the previous section is said to be an *integral domain*. (If axiom 7 is omitted, it is called a *ring*.) Thus to stipulate that the integers (together with their arithmetical operations of addition and multiplication) satisfy these axioms, it would be enough to say that they constitute an integral domain. The rational numbers (fractions), the real numbers, and the complex numbers are other examples of integral domains.

Any result proved by means of logical argument starting from a given axiom system will be "true" for the abstract structures that satisfy that axiom system, but questions about the "truth" of the result in the real world would not just be unanswerable, *they would have no meaning.* If the axioms are a good reflection of some phenomenon in the real world, the consequences of those axioms will also relate well to the real world and may even provide some useful information that can benefit (or possibly lead to the destruction of) the human race. But as far as the mathematics is concerned, the relevance or otherwise of the initial axioms is immaterial. Some axiom systems that have led to very interesting mathematics appear to have no relation whatsoever to the physical world—although this is not to say that a connection will not one day be discovered. At the price of isolating themselves from reality (to a greater or lesser extent), mathematicians are able to work in a world of absolute certainty, with the potential bonus of having their results find widespread application (within mathematics itself in the first instance) because their axiom systems fit structures other than the ones they (may have) had in mind.

But if "truth" cannot be the guide, what are the considerations that do govern the formulation of an axiom system?

An essential requirement is *consistency:* it must not be possible to deduce two mutually contradictory consequences from the axioms. This requirement must be satisfied by all axiom systems, although proving consistency is not only difficult but fraught with philosophical problems as well (as will be indicated presently).

The other requirement, which is pertinent to any axiom system that attempts to describe some particular mathematical structure (such as the arithmetic of the integers), is *completeness.* The axiom system should be rich enough to enable all the "true facts" about the structure to be provable.

Satisfying both these requirements is a delicate balancing act. To achieve completeness, more and more axioms may need to be introduced. And yet the more axioms there are, the greater the likelihood will be of introducing an inconsistency.

The Gödel Incompleteness Theorems

At the turn of the century, the world-famous German mathematician David Hilbert proposed a program for the development of all mathematics within the strict formalization of the axiomatic method. According to Hilbert, all mathematics could be regarded as the formal, logical manipulation of symbols based on prescribed axioms. (This would mean that in principle, a computer could be programmed to "do all of mathematics.") But in 1930, with two startling and totally unexpected theorems, the young Austrian mathematician Kurt Gödel demonstrated that Hilbert's program could not possibly succeed.

Gödel proved that for any consistent axiom system strong enough to allow for the development of elementary integer arithmetic, there will always be statements relevant to that axiom system that can be neither proved nor disproved based on those axioms (first incompleteness theorem). Moreover, among those unprovable (from the axioms) statements is the statement that the axiom system is consistent—so the all-important (for Hilbert's program) notion of consistency is destined to remain forever elusive (second incompleteness theorem).

Although Gödel's results meant that the axiomatic method could not be elevated to the all-embracing position envisaged by Hilbert, it should not be thought that it heralded the death of the axiomatic approach in everyday mathematics as it was, and still is, practiced. On the contrary, the twentieth century saw the axiomatic method assume a supreme position in mathematics. Instead, Gödel forced us to abandon the belief or hope that an axiom system could be adequate to answer all the questions we might reasonably ask of it.

In fact, with the increasing success of the axiomatic approach, the years after Gödel's pronouncement saw a gradual growth in the belief that it was only highly specialized statements that could not be proved. For instance, Gödel's first incompleteness theorem was obtained by showing that in elementary number theory, it is possible to formulate a statement analogous to the—in English—obviously paradoxical statement

> *The sentence enclosed in a box on this page is false.*

(In Gödel's number-theoretic analogue, "false" is replaced by "not provable." It is because elementary arithmetic is necessary to formulate such a statement that Gödel's result applies only to axiom systems that provide this facility. But of course, any axiom system destined to fulfill the aims of the Hilbert program would certainly enable you to perform elementary arithmetic.)

Also, in regard to Gödel's second incompleteness theorem, although the consistency of an axiom system is an important consideration, the fact that it cannot be proved from the axioms is not such a great problem. Writing down the axioms in the first place entails a tacit assumption of their consistency, and the main interest lies in the more "solid" consequences of the axioms. For instance, in number theory, the statement that the axioms are consistent is not the kind that number theorists are generally concerned about. So perhaps Gödel's incompleteness results are not so relevant after all. Or so it once seemed before a brash young American came along and proved otherwise. Paul Cohen's devastating 1963 result completely shattered the cozy feeling that incompleteness did not affect "real" problems, and it did so in the most basic and fundamental part of mathematics: *set theory.*

Axiomatic Set Theory

At the turn of the century, the development of abstract, pure mathematics (and, in particular, the various subjects stemming from Newton's and Leibniz's *infinitesimal calculus*) that had taken place during the nineteenth century led the German mathematician Georg Ferdinand Ludwig Philipp Cantor to formulate a very general mathematical "framework" that could serve as a foundation for all mathematics. The subject created by Cantor is still with us and serving as a good foundation. It is known as *set theory,* and its concepts and methods pervade nearly all present-day mathematics. But its development since Cantor's original formulation has been both violent and traumatic, as the following pages will reveal. Before that, however, I should say a little about formal logic.

At the same time as Cantor was developing his ideas of set theory and, in particular, a system of infinite numbers to measure the "size" of infinite sets, Gottlob Frege was creating what is now known as *predicate logic.* Broadly speaking, this provides a universal, formalized language adequate for expressing any mathematical concept whatsoever. It was not that the importance of this development stemmed from any great need or desire by mathematicians to carry out their work using predicate logic. Indeed, owing to the simplicity of the language, in most cases the expression of a mathematical concept or argument within Frege's framework is extremely long and cumbersome. Rather, Frege's work was important, first, because it clearly demonstrated that all the many branches of mathematics are part of a single, coherent whole and, second, (and far more important) because it enabled mathematicians to analyze the deductive methods they used when constructing proofs. (I should note that recently predicate logic has been used for the actual expression of mathematical concepts and proofs, in attempts to develop computer programs to obtain, or help obtain, mathematical results. To present mathematics in a form suitable for a computer to handle, a precise and simple language must be used, and predicate logic provides just such a framework.)

The concept of a set as used by Cantor was extremely simple. A *set* is any collection of objects, or at least any collection of mathematical objects. The key is to regard the collection as *a single object in its own right*. Small, finite sets can be described by listing their *members* (or *elements*), usually enclosing the list in curly brackets. Thus

$$\{1, 3, 5, 9\}$$

denotes the set whose elements are the numbers 1, 3, 5, and 9. For larger (and possibly infinite) sets, it is not possible to list all the elements, and so you must rely on a *property* to determine what set you are thinking of. The standard notation for the set of all those objects x for which the property $P(x)$ is valid is

$$\{x \mid P(x)\}.$$

Thus, the set of all prime numbers (an infinite set) may be denoted by

$$\{x \mid x \text{ is a prime number}\}.$$

Some sets have no defining property, and such sets cannot be specified in terms of their elements, but this point is not really relevant to our elementary discussion. Roughly speaking, such sets arise "by default," since the notion of a collection does not entail the existence of a property that determines it—but this is an advanced, subtle point. Ignoring these elusive sets in favor of those determined by properties, what kind of "properties" are allowed in the formation of sets? The original answer was, as you might expect by now, any property that can be expressed in Frege's predicate logic, a definition that by the very nature of this formalized language is both precise and adequate for all the properties encountered in mathematics.

At this stage, things could hardly seem more rosy. Set theory provides an adequate foundational framework on which all mathematical objects and structures may be constructed, and Frege's predicate logic provides a universal language for defining and discussing these objects and structures, including the basic notion of a set itself. Frege himself used set-theoretic concepts extensively in his two-volume work *The Foundations of Arithmetic,* intended as the culmination of his entire life's work.

It was while the second volume of Frege's book was at the printers that he received a letter dated June 16, 1902, from the famous British logician Bertrand Russell. After an initial paragraph praising Frege's first volume, Russell came to the purpose of his letter. "There is just one point where I have encountered a difficulty," he began. Then he briefly explained an observation he had made exactly one year earlier—an observation that completely destroyed Frege's entire theory.

Russell's paradox, as it is known, is as simple as it is profound. According to the basic principle of Cantor's set theory, if $P(x)$ is any property (expressible in predicate logic) applicable to the mathematical object x, then there is a corresponding set of all those x for which $P(x)$ is true, that is, the set

$$\{x \mid P(x)\}.$$

There is nothing to stop the objects x involved here from being sets themselves, for a set is a mathematical object just like any other. (Indeed, when set theory is used as the basic foundation of mathematics, every mathematical object is a set of some kind.) Russell now took for the property P the statement (applicable to sets x)

$$R(x): x \text{ is not a member of } x.$$

(The standard symbol for set membership is $\in$, so $x \in y$ means that x is a member of y, and nonmembership is denoted by $x \notin y$, so Russell's property $R(x)$ may be written as $x \notin x$.)

Now let us give a name to the set determined by property $R(x)$, say y. Thus

$$y = \{x \mid x \notin x\}.$$

Since y is a set, it is reasonable to ask whether y is a member of itself. If it is, then y must satisfy its own defining property, which means that $y \notin y$—that is, y is *not* a member of itself. On the other hand, if y is not a member of itself, then y cannot satisfy its defining property, so $y \in y$ must hold—y is a member of itself. Accordingly, we have arrived at the obviously untenable situation in which if y is a member of itself, then it is not; and if it is not a member of itself, then it is. A genuine paradox.

What made the Russell paradox so devastating was its simplicity. It employed only the most basic concepts, concepts on which almost all mathematics depends.

A way out of the dilemma caused by Russell's paradox was provided by the German mathematician Ernst Zermelo, whose work on integral equations (a highly applicable area of mathematics) had led him to consider deep problems concerning the nature of infinite sets. To establish a secure set-theoretical framework for his work, he published a paper in 1908 in which he developed a system of axioms for set theory. Subsequently modified by Abraham Fraenkel, *Zermelo-Fraenkel set theory* gradually came to be accepted as the "correct" axiomatic approach to the theory of abstract sets. (An adequate motivation and explanation of the axioms involved requires more space than is available here, although an elementary account can be found in various texts—details later.)

According to Gödel's incompleteness theorems, there is, of course, no possibility of *proving* that the axioms of Zermelo-Fraenkel set theory are consistent, but they certainly do appear to avoid paradoxes such as Russell's paradox, and most mathematicians believe that they will not lead to any contradiction at all—a belief that has grown stronger as the theory has shown that it can stand the test of time and heavy usage.

So much for consistency. What about completeness? The incompleteness theorems also tell us that there will be statements about sets that can be neither proved nor disproved on the basis of the axioms adopted. This deficiency assumes a greater importance than usual because of the special, foundational nature of set theory. Since the entire edifice of modern mathematics can be regarded as (and, to a great extent, explicitly is) built on set theory, deficiencies in set theory might result in genuine deficiencies in other areas of mathematics. But although this was always a possibility, the Zermelo-Fraenkel axioms did appear to be adequate to provide a theory of sets sufficient for mathematics, and most working mathematicians quietly ignored this danger, assuming that it would not affect them. Until, that is, Cohen forced the issue out into the open with his breakthrough in 1963.

Although Cohen's discovery turned out to have many ramifications, it was initially concerned with a problem involving Cantor's infinite numbers, the theory of which became perfectly sound (as far as anyone knows) once the Zermelo-Fraenkel axioms had been formulated. So now we should journey into the infinite and take a look at Cantor's theory.

Infinite Sets

Even though the world we live in is finite, the mathematics we need in order to deal with it involves the infinite at almost every turn: the set of all natural numbers is an infinite set; the precise specification of the number π requires infinitely many decimal places; the number of points on the smallest of lines is infinite; and so on. Although attempts have been made to avoid all use of the infinite, the resulting mathematics is incredibly cumbersome and unwieldy. For despite its total abstraction, the infinite is a world of great simplicity. Going from the finite to the infinite is very much like stepping back from a television screen: when you are far enough away, the indecipherable complexity of the large number of tiny light spots occupying the screen is recognized as a coherent picture. By going to the infinite, the complexity of the large finite is lost. This is a phenomenon not restricted to pure mathematics alone. In economics, for example, idealized economies with infinitely many traders are studied in preference to the very large finite economies of the real world, and in physics, infinite volumes are used when discussing certain subtle concepts of heat and electrical energy.

It was the development of a system of infinite numbers and their arithmetic that formed the crowning achievement of Cantor's pioneering work on set theory. But why do we need infinite numbers at all? you might ask. The answer is, For the same reason that we need finite (whole) numbers—to count the "number" of members of a set. The natural numbers enable us to "measure the size" of a finite set. To measure the size of an infinite set, infinite numbers are required (from which remark you can anticipate that it is not enough to refer to such a set simply as *infinite*). Having accepted that point, you might then ask, What is an infinite number? To this question, a good response would be, What *is* a finite number? As we saw at the beginning of this chapter, natural numbers are mere figments of our imagination, so postulating the existence of infinite numbers should be no different. What is important is the way these infinite numbers behave, and this is where the key to Cantor's infinite numbers lies.

Natural numbers are abstracted from finite sets (either mathematical ones or real-life sets like sets of apples, sets of people, and so on). The number three is what all sets of three elements have in common. On the face of it, this looks like a circular definition (which thus would not be a definition at all, of course), but Cantor observed that this was not the case. Rather, before we define number, we must understand the notion of "same size" for two sets, as will now be explained.

Two sets, call them *A* and *B,* have the *same size* if their elements can be matched in such a way that each member of *A* is matched to exactly one member of *B,* and vice versa. Thus, for example, the sets

$$A = \{1, 2, 3, 4\}, \quad B = \{100, \pi, \sqrt{2}\tfrac{1}{2},\}$$

are the same size, as is shown by the matching (there are others):

$$
\begin{array}{cccc}
1 & 2 & 3 & 4 \\
\updownarrow & \updownarrow & \updownarrow & \updownarrow \\
100 & \pi & \sqrt{2} & \tfrac{1}{2}
\end{array}
$$

Similarly, the sets

$$A = \{a, b, c\}, \quad B = \{\text{foot, sock, shoe}\}$$

are the same size because of the matching

$$
\begin{array}{ccc}
a & b & c \\
\updownarrow & \updownarrow & \updownarrow \\
\text{foot} & \text{sock} & \text{shoe}
\end{array}
$$

Notice that in neither case does the concept of the number of elements in the set arise. In order to talk about "same size," it is not necessary to have a prior notion of "size," nor is there any need to consider only finite sets. The same ideas apply to infinite sets (although in this case, it is not possible to describe the matching explicitly). When these ideas are applied to infinite sets, however, you soon find yourself faced with some unexpected results. For example, let A be the set of all natural numbers, and let B be the set of all even natural numbers. Intuitively, B is exactly "half the size" of A, but according to our definition, these two sets are the same size, as shown by the matching

$$
\begin{array}{cccccc}
1 & 2 & 3 & 4 & 5 & \ldots \\
\updownarrow & \updownarrow & \updownarrow & \updownarrow & \updownarrow & \updownarrow \\
2 & 4 & 6 & 8 & 10 & \ldots
\end{array}
$$

There is, however, no contradiction here—or if there is, it is only with our preconceptions. It is just that infinite sets do not always behave in the same way as finite sets do.

A nice illustration of the kind of behavior that arises with infinite sets is provided by *Hilbert's hotel*. This imaginary institution has an infinite number of rooms, numbered 1, 2, 3, and so on all the way through the natural numbers. One night, as chance would have it, all the rooms are taken. (In this story, there are infinitely many people as well.) And yet a latecomer can still be accommodated, without anyone's having to be thrown out. All that needs to be done is to put the newcomer into room 1, moving the present occupant of that room into room 2, that room's occupant into room 3, and so on. All the guests already there are moved one room along, allowing the latecomer to move into the now empty room 1. (In fact, it is possible to accommodate infinitely many latecomers. Can you see how?) Although the idea of an infinite hotel might seem far-fetched, there is nothing wrong with the internal logic of the discussion. However counterintuitive, this is the kind of thing that happens when you start to explore the world of the infinite.

The example of the natural numbers and the even natural numbers might lead you to suspect that all infinite sets are the same size, which would mean that there is no need for a system of infinite numbers. Indeed, many of the infinite sets commonly encountered in mathematics are the same size. For instance, the set of prime numbers, the set of natural numbers, the set of integers, and the set of rational numbers all have the same size. (Sets having the same size as the natural numbers are often referred to as *countable*, since matching them with the natural numbers provides a way of counting out their elements.) But as Cantor discovered, not all infinite sets are the same size. In fact, there is a whole (infinite) hierarchy of infinities, getting larger and larger all the while. Cantor's proof of this key fact is both simple and elegant, using only the most basic notions of set theory, but it is highly abstract and, for this reason, will be left until the very end of the chapter, where the more squeamish readers may skip it. I mention at this point only that the set of real numbers is not the same size as the set of natural numbers (although it is the same size as the set of points in the plane and the set of points in three-dimensional space).

Infinite Numbers and Cantor's Continuum Problem

Once you have grasped the concept of same size, you can develop a system of "numbers" that may be used to measure the "size" of any set, be it finite or infinite. The "numbers" themselves will just be abstractions, of course. The important point is that if two sets have the same size (i.e., if their elements can be paired off, as described in the last section), their "size" (i.e., the "number" of elements in each set) should be the same. So, for instance, when you measure the "size" of the two sets

$$\{a, b, c\}, \quad \{\text{Fred, Elsie, Fido}\},$$

you find that both sets have the same "number" of elements, namely, three. Likewise, when you measure the "size" of the two infinite sets

$$\{1, 2, 3, 4, 5, \ldots\}, \quad \{2, 4, 6, 8, 10, \ldots\},$$

you find again that these sets have the same "number" of elements—in this case, this "number" is the smallest of the infinite numbers, denoted (following Cantor) by the symbol $\aleph_0$. (This is read as "aleph-null"; aleph is the first letter of the Hebrew alphabet. The reason for the subscript zero will become clear in a moment.)

So what *is* the "number three"? It is what all sets of three elements have in common. Or to put it in another way, it is what is common to all sets having the same size as the set $\{a, b, c\}$. Thus 3 is an abstraction that comes out of the notion of same size. There are various mathematical ways of making this statement precise, none of which we will explore here. The main point is that finiteness is not relevant. So if you are happy (*were* happy?) with the concept of "the number three," you should have no less ease in accepting the "number" $\aleph_0$. It is what is common to all sets having the same size as the set of all positive whole numbers.

As mentioned earlier, not all infinite sets have the same size—there is a whole (infinite) hierarchy of infinities. So, just as there is an infinite list of finite numbers, 1, 2, 3, . . . , so too there is an infinite list of infinite numbers, $\aleph_0$, $\aleph_1$, $\aleph_2$, $\aleph_3$, . . . , each one "bigger" than the one before.

The addition and multiplication of Cantor's alephs turn out to be particularly simple (if, at first sight, a bit surprising). In each case, the result is just the larger of the two infinite numbers. For example,

$$\aleph_0 + \aleph_1 = \aleph_1,$$

$$\aleph_1 + \aleph_3 = \aleph_3,$$

(Hilbert's hotel corresponds to the fact that $\aleph_0 + 1 = \aleph_0$. In order to cause an overflow from the hotel, $\aleph_1$ guests would have to arrive.)

Many of the infinite sets that occur in mathematics have size $\aleph_0$. For instance, the set of positive whole numbers, the set of all (i.e., positive and negative) whole numbers, the set of all rationals, and the set of all prime numbers all have size $\aleph_0$. But as Cantor showed, the set of all real numbers definitely has more than $\aleph_0$ members. This at once raises the question, Just what is the size of this set? Since it is not $\aleph_0$, it must be one of $\aleph_1$, $\aleph_2$, $\aleph_3$ . . . , but which one? Despite many attempts, Cantor himself was unable to answer this seemingly simple question, and so too were a number of other excellent mathematicians. Indeed, *Cantor's continuum problem,* as the question became known, resisted so many attempts at solution that when Hilbert gave the keynote address at the International Congress of Mathematicians in Paris in 1900, he included the problem in a list of what he saw as the most significant challenges facing mathematicians at the dawn of the new century. It is called the continuum problem because it asks for the size of the real *continuum*— the word used to describe the set of real numbers when considered as the points that make up the *real line*.)

Some progress was made in 1938, when Kurt Gödel used new techniques of mathematical logic to show that based on the Zermelo-Fraenkel axioms, it is definitely not possible to prove that the set of real numbers does not have size $\aleph_1$. But this did not solve the problem, since it was still possible that the axioms were simply not sufficient to decide the issue one way or the other.

Despite this possibility, however, in the years following Gödel's result, it was widely expected that the continuum problem could be decided within the Zermelo-Fraenkel framework. In this case, since Gödel had shown that the answer definitely could not be proved to be anything other than $\aleph_1$, it had (it was assumed) to be the case that the continuum did have size $\aleph_1$, and given time, this would eventually be proved conclusively. Accordingly, it did not seem unreasonable to assume this anticipated result whenever a piece of mathematics required knowledge of the size of the continuum, and many results were proved on the assumption that the *continuum hypothesis* (as the assumption was known) was true.

Then in 1963 came the announcement that Paul Cohen of Stanford University had developed a new logical technique by which he had been able to *prove* that the continuum hypothesis could not be deduced from the Zermelo-Fraenkel axioms. When combined with Gödel's earlier result, this showed that the continuum hypothesis could not, in fact, be decided within the Zermelo-Fraenkel system.

So what next? There are two ways of regarding the state of affairs resulting from Cohen's result. One conclusion is that it demonstrates the inadequacy of Zermelo and Fraenkel's axioms. According to this viewpoint, the inadequacy is clearly very severe. It is one thing to know that the system is deficient, as predicted by Gödel's incompleteness theorems, but for the system to be unable to resolve such a basic question as How many real numbers are there? strikes a devastating blow at the theory. Some mathematicians reacted to this blow by suggesting that additional axioms should be formulated to overcome the newfound deficiency. But if you take that route, you are faced with the task of having to find appropriate additional axioms. Because of the essentially simple nature of set theory and its foundational position in mathematics, any axioms you introduce will have to be "believable"—which means that even if they are not immediately obvious (and some of the Zermelo-Fraenkel axioms require some thought in order to see what they are saying), they will have to appear natural once you start to study them. (This consideration prevents you from taking the easy way out, of simply adopting the continuum hypothesis itself as an axiom of set theory—what justification is there for doing so?) The fact that mathematicians have been working with axiomatic set theory for almost a century now without encountering any such additional "principle" leads most experts to conclude that there is, in fact, no "missing axiom."

So what you are left with is the alternative conclusion to be drawn from Cohen's discovery: that however unpalatable it may seem, there simply is not one set theory, but several. (Likewise, the nineteenth century brought the realization that there is not a single "correct" geometry, but three, alternative, geometries, each with its own distinctive properties and results.) In some set theories, the continuum hypothesis is true; in others, it is false.

Did you notice my use of the phrase "several set theories," not just "two"? It is not just the continuum hypothesis that cannot be decided from the Zermelo-Fraenkel axioms. Following Cohen's initial discovery in 1963, it became apparent that his new method (the method of *forcing*) was applicable in many situations and not only in set theory itself. In the following two decades the undecidability of many classical unsolved problems of mathematics was demonstrated. Gone forever was the old expectation that given enough time and ingenuity, any "genuine" problem of mathematics could be resolved one way or the other. Besides the true statements and the false statements is a third class of *undecidable* statements, statements that are neither true nor false. But at least Cohen's method of forcing provided a means of demonstrating that a problem belonged to this third class, so his result did make a positive contribution to mathematics.

Cantor's Proof

How did Cantor prove that there is a whole hierarchy of infinities? For readers who would like to see an example of highly abstract mathematical reasoning, the following is a modernized version of Cantor's argument. It starts with the set-theoretic notion of a *subset*.

If X is any set, then any collection of objects taken from X is called a *subset* of X. Thus, the collection

$$\{a, c, d\}$$

is a subset of the collection

$$\{a, b, c, d, e, f\},$$

and the set of prime numbers is a subset of the set of all integers.

Consider now the set consisting of *all* the subsets of the set X. Does such a set exist? Memories of the Russell paradox should be enough to warn you that you must be careful about postulating the existence of sets. In this case, there is no problem (as far as anyone knows). One of the axioms of Zermelo-Fraenkel set theory guarantees that there is such a set: it is called the *power set* of X and is denoted by $\mathcal{P}(X)$. For example, if $X = \{a, b\}$, then $\mathcal{P}(X)$ consists of the sets

$$\varnothing, \{a\}, \{b\}, \{a, b\}.$$

What is the symbol $\varnothing$? It denotes the *empty set* (or *null set*), the set with no members. *If* it is indeed a set, it will surely be a subset of any other set, since in a trivial but nonetheless logically valid way, a set with no members has the property that "all its elements" lie in any set X you choose. On the basis of this reasoning alone, you may be tempted to think that it is not a good idea to include among the other "fictions" of mathematics the notion of an empty set. But then, by the same token, zero is not a sensible number. And here you have some idea why the empty set *is* considered a bona fide set, along with all the other sets in mathematics. It is a "zero element" much like the number 0. (Indeed, 0 is the number of elements in the set $\varnothing$.)

Another example: if $X = \{1, 2, 3\}$, then $\mathcal{P}(X)$ is the set whose members are the sets

$$\varnothing, \{1\}, \{2\}, \{3\}, \{1, 2\}, \{1, 3\}, \{2, 3\}, \{1, 2, 3\}.$$

These two examples should be enough to indicate that $\mathcal{P}(X)$ seems to be a much larger set than X. When X has two members, $\mathcal{P}(X)$ has four, and when X has three members, $\mathcal{P}(X)$ has eight. In fact, a fairly straightforward mathematical proof shows that if a finite set X has n members, then $\mathcal{P}(X)$ has 2^n elements. You will appreciate from our discussion of the exponential function 2^n in chapter 1 that the size of $\mathcal{P}(X)$

increases very rapidly as more elements are added to X. It turns out that this difference in growth rates carries over into the infinite (although as Hilbert's hotel demonstrated, such carryovers from the finite to the infinite should never be taken for granted). This is how Cantor proved that there are infinitely many infinities. He demonstrated that for any set X, $\mathcal{P}(X)$ is definitely not the same size as X (and hence is of a greater order of magnitude than X). The infinitude of the infinities follows easily from this, as follows. Suppose that X is the set of natural numbers. Then $X_1 = \mathcal{P}(X)$ is a set of greater infinite magnitude than X. Again, $X_2 = \mathcal{P}(X_1)$ is larger than X_1. Similarly, $X_3 = \mathcal{P}(X_2)$ is bigger than X_2, and so on.

To prove Cantor's result, suppose for the sake of argument that $\mathcal{P}(X)$ were the same size as X. Our aim is to deduce a contradiction from this assumption. Then, provided the argument used is logically sound, the inescapable conclusion will be that the initial assumption must be false (since a true assumption cannot lead to a false or contradictory result). So here goes: Since X and $\mathcal{P}(X)$ are assumed to be the same size, there will be a matching between these two sets. That is, for each member x of X, there will be an associated member "partner-of-x" in $\mathcal{P}(X)$ that is not matched to any other member of X; moreover, every member of $\mathcal{P}(X)$ will be the partner of some member of X. Because this argument is intended to be valid for any set X, finite or infinite, there is no possibility of representing this matching using arrows, as we did earlier—but see box A.

Now consider an arbitrary member x of X. Its partner, call it A, is a member of $\mathcal{P}(X)$; that is, A is a subset of X. So A consists of some of the members of X. Is x itself among these elements; that is, is x a member of A? This is a perfectly reasonable question. For some members x of X, the answer presumably is yes; for others, no. Let U be the set of all those members x of X for which x is not a member of partner-of-x. (Does this remind you of anything?) The set U consists of elements of X, and hence U is a subset of X; that is, U is a member of $\mathcal{P}(X)$. Thus U is the partner of some member of X, call it w. (So $U =$ partner-of-w.)

Now ask, Is w a member of U? If it is, then w must satisfy the requirement defining U, which is to say that w is not a member of its partner (in this case, U). On the other hand, if w is not a member of U, then w fails to satisfy the requirement defining U, so w will be a member of its partner, U. This is an untenable situation—a contradiction. As mentioned earlier, the only possible conclusion is that the initial assumption

Box A

Proof of Cantor's Theorem

Given an infinite set X, the proof demonstrates that $\mathcal{P}(X)$ does not have the same size as X (and hence is bigger). The idea is to show that the supposed existence of a matching between the elements of X and those of $\mathcal{P}(X)$ leads to a contradiction. If the members of X are denoted by letters of the alphabet (a finite set, of course, but good enough for this illustration), then part of such a supposed matching might look like this:

Element of X		Subset of X
y	$\leftrightarrow$	$\{a, b, c, d\} = A(y)$
z	$\leftrightarrow$	$\{b\, d, p, q, z\} = A(z)$

In this case, y is not a member of the set $A(y)$ with which it is matched, whereas z is a member of its matched set $A(z)$. Let U denote the set of all those members of X that are not members of their matched set:

$$U = \{x \mid x \notin A(x)\}.$$

Since U is a subset of X, U must be matched with some element of X, say w:

$$w \qquad \leftrightarrow \qquad U = A(w).$$

Investigation of whether or not w is a member of the set $A(w)$ now leads to a contradiction. If w is a member of $A(w)$, this means that w cannot be in U, but since U is just $A(w)$, this is contradictory. But if w is not a member of $A(w)$, then w will be a member of U, and again, since $U = A(w)$, this is a contradiction.

that X and $\mathcal{P}(X)$ were the same size had to have been false. Thus X and $\mathcal{P}(X)$ cannot be the same size, and Cantor's result is proved.

This is an excellent example of how a disaster can be turned to great advantage. The similarity between the argument just given and Russell's paradox cannot have escaped you. In this case, however, all the steps can be justified according to the Zermelo-Fraenkel axioms, and instead of the contradiction resulting in a paradox that destroys the entire theory, what you get is the desired falsity of the initial supposition.

And there you have it. The development of set theory has been one of the great success stories of modern mathematics, and there is nowadays scarcely any branch of mathematics that is not influenced to a greater or lesser extent by set-theoretic ideas and methods. Bertrand Russell once praised Cantor's achievement in giving birth to this subject as "possibly the greatest of which the age can boast," and David Hilbert claimed, "From the paradise created for us by Cantor, no one will drive us out." Few would argue with these opinions.

Suggested Further Reading

For an introductory account of mathematical logic, read *What Is Mathematical Logic?* by C. J. Ash, C. J. Brickhell, et al. (Dover, 1990) or *Introduction to Mathematical Logic* (4th edition), by Elliott Mendelson (Thomson, 1997).

A gentle but thorough introduction to set theory that covers all the topics mentioned in this chapter is *The Joy of Sets,* by Keith Devlin (Springer-Verlag, 1992).

3

Number Systems and the Class Number Problem

The Solution to a 180-Year-Old Problem

In 1983, Don Zagier of the University of Maryland and the Max Planck Institute in Bonn, and Benedict Gross of Brown University, Providence, Rhode Island, announced that they had solved the class number problem, a famous (among mathematicians) problem posed by Gauss in 1801. Although their proof was by no means the longest in mathematics (chapter 5 deals with that), at 300 pages it was longer than most. But it is not the length of the proof that mathematicians found so fascinating; it was its nature. It was very indirect and linked two seemingly unrelated areas of mathematics in quite a remarkable way.

Although both the problem and its solution are highly abstract and involve some very difficult mathematics, at heart it concerns number systems of one kind or another, and it certainly is possible to describe the general issues entailed. This is what this chapter sets out to do. Along the way we shall trace out much of the historical development of present-day mathematics. But we start out by taking a look at the following.

The Remarkable Properties of the Number 163

In the eighteenth century, the great Swiss mathematician Leonhard Euler discovered (no one knows how) that the formula

$$f(n) = n^2 + n + 41$$

has a rather remarkable property. If you make n equal to any of the numbers from 0 to 39, the resulting value of $f(n)$ is a prime number. For instance, $f(0) = 41$ is prime, as are $f(1) = 43$ and $f(2) = 47$. No other quadratic formula has been discovered that produces anything like as many prime numbers (starting from $n = 0$ and working up through successive values of n). And although the unbroken sequence of prime numbers stops with $n = 40$, when $f(40) = 41^2$, the formula still produces a lot of prime numbers. Of the first 10 million values, the proportion of primes is about one in three—far greater than for any other quadratic formula (see chapter 6 for more discussion of prime-generating formulas).

Since Euler's formula seems so unusual in its production of primes, there is presumably something rather special about it. But what? Well, properties of formulas for integers often turn out to be closely related to properties of the same formulas regarded as formulas for real numbers (or even for complex numbers). Indeed, an entire branch of mathematics, known as *analytic number theory*, exploits this phenomenon (see chapter 9). What happens when Euler's formula is regarded as a formula for real numbers?

First, let us rewrite the formula with x (the usual symbol for a real number) in place of n (the usual symbol for an integer), thus

$$f(x) = x^2 + x + 41.$$

Those readers who can remember the algebra they learned at school ought to find this reminiscent of *quadratic equations*—equations of the form

$$ax^2 + bx + c = 0$$

—which must be solved for x when the values of a, b, and c are known. You may even remember the formula that gives the two solutions:

$$x = \frac{-b \pm \sqrt{b^2 - 4ac}}{2a}$$

(The two possibilities for the "plus-or-minus" sign give two solutions to the equation.)

Because it is not possible to take the square root of a negative number (when real numbers are being considered), the sign of the expression $b^2 - 4ac$ is very important. If this expression is positive, the quadratic equation will have two solutions; if it is negative, there will be no (real) solutions. (And if it is zero, then there will be only one solution—this is a special case.) The expression is called the *discriminant* of the quadratic equation.

What is the discriminant of the Euler quadratic

$$x^2 + x + 41?$$

Here $a = 1$, $b = 1$, and $c = 41$, so

$$b^2 - 4ac = 1 - 164 = -163.$$

Because the discriminant is negative, we know right away that the quadratic equation

$$x^2 + x + 41 = 0$$

has no (real) solutions. (There are two complex solutions:

$$x = -\tfrac{1}{2} \pm \tfrac{1}{2}\sqrt{163}i.$$

See later in this chapter for a discussion of complex numbers.)

And there, believe it or not, lies the reason for the special behavior of the Euler formula as a prime generator. The reason is not that the discriminant is negative—lots of formulas have that property—but that its value is -163. What's so special about 163? you might ask. Read on, and you will discover that it is a very special number indeed, closely related to some fundamental mathematical constants.

What are the most common special "constants" of mathematics, the numbers that keep cropping up in the most unexpected places? The most obvious one is π, the ratio of the circumference of a circle to its diameter. (This definition already indicates that π is special. Why should you get the same answer for every circle, whatever its size?)

Since the late eighteenth century, it has been known that π is *irrational*, which is to say that its decimal representation continues indefinitely, without settling into any repetitive pattern. To twenty decimal places,

$$\pi = 3.14159265358979323846.$$

Computers have been used to calculate π to more than 30 million places.

Besides its definition in terms of circles, π occurs in many other situations. For instance, the sum of the terms in the infinite sequence

$$1 - \frac{1}{3} + \frac{1}{5} - \frac{1}{7} + \frac{1}{9} - \frac{1}{11} + \ldots$$

has the value $\pi/4$. (The development of methods for dealing with infinite sums such as this was one of the crowning achievements of nineteenth-century mathematics.) Again, the sum of

$$1 + \frac{1}{4} + \frac{1}{9} + \frac{1}{16} + \frac{1}{25} + \ldots$$

(where the nth term in the sequence is the reciprocal of n^2) is $\pi^2/6$.

Another surprising appearance of π is this: if you throw a matchstick onto a board on which are ruled parallel lines one matchstick-length apart, the probability that the matchstick will end up touching one of the lines is precisely $2/\pi$.

After π, the next most common mathematical constant is e, the base of the natural logarithms. Like π, the number e is irrational. To twenty decimal places,

$$e = 2.71828182845904523536.$$

And like π, there are various ways of defining e. For instance, e is the number for which the graph of the function

$$y = e^x$$

has the property that the slope at any point equals the value (of y) at that point. For example, if a population p is growing according to the law

$$p = e^t$$

(where t is the time), the rate of growth at any instant is exactly equal to the population size at that instant.

Another (related) definition of e is that it is the number for which the area bounded by the lines $y = 1/x$, $y = 0$, $x = 1$, and $x = e$ is exactly equal to 1 (see figure 3). Expressed in terms of an integral, this definition of e gives the equation

$$\int_1^e \frac{1}{x}\, dx = 1.$$

Still another definition involves an infinite sum:

$$e = 1 + \frac{1}{1!} + \frac{1}{2!} + \frac{1}{3!} + \frac{1}{4!} + \ldots,$$

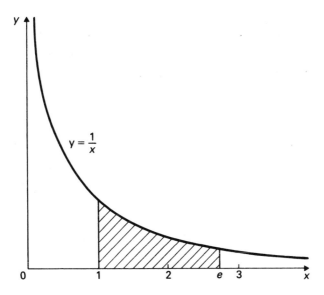

FIGURE 3   The definition of the mathematical constant e as the number for which the shaded area is exactly 1.

where $N!$ (to be read as "N factorial") denotes the product $1 \times 2 \times 3 \times \ldots \times N$. In fact, this is just a special case of the formula

$$e^x = 1 + \frac{x}{1!} + \frac{x^2}{2!} + \frac{x^3}{3!} + \frac{x^4}{4!} + \ldots .$$

Anticipating for a moment the topic of complex numbers to be discussed later in this chapter, the preceding formula is valid even if the number x is *complex*, that is, if it has the form $a + bi$, where $i = \sqrt{-1}$. This leads to some amazing results. For example, Euler discovered that

$$e^{\pi i} = -1.$$

In other words, when the irrational number e is raised to the power of the irrational number π times the imaginary number $\sqrt{-1}$, the result is the whole number -1. Another equally startling result that relates e, π, and $\sqrt{-1}$ is

$$i^i = e^{-\pi/2} = 0.2078795763 \ldots .$$

And now to the point of this discussion of mathematical constants: The three numbers π, e, and $\sqrt{163}$ are all irrational. And yet, to twelve decimal places,

$$e^{\pi\sqrt{163}} = 262{,}537{,}412{,}640{,}768{,}744.000000000000.$$

In fact, this number is not an integer. A more accurate value is

$$262{,}537{,}412{,}640{,}768{,}743.999999999999250,$$

which is correct to fifteen decimal places. Nevertheless, the result is very nearly an integer, whereas for most other natural numbers k, the value of $e^{\pi\sqrt{k}}$ is not nearly so close. The surprising thing is that the value of k for which this occurs is again the number 163. You might suspect that this is not just an accident but that there is something going on behind the scenes, and you would be right. Just what is so special about this number 163 will be revealed as the remainder of this chapter unfolds. The story begins in ancient Greece.

Early Number Systems

The ancient Greeks appear to have been the first to develop a mathematical theory of arithmetic. Both the Ionian school (founded by Thales around 600 B.C.) and the Pythagorean school (founded by Pythagoras some fifty years later) developed extensive theories of both geometry and (particularly the Pythagoreans) arithmetic. It was the Greeks who first recognized the *natural* (or *counting*) *numbers* 1, 2, 3, . . . as forming an infinite collection on which the basic arithmetic operations of addition and multiplication could be performed. Although they did not acknowledge *negative* numbers as such, they knew how to manipulate minus signs in expressions like

$$(7 - 2) \times (6 - 3) = (7 \times 6) - (7 \times 3) - (2 \times 6) + (2 \times 3).$$

In fact, their approach was probably not unlike that summed up by the old schoolroom jingle:

Minus times minus equals plus,
The reason for this we need not discuss.

However, there was a good reason that the Greeks refused to countenance as a number an entity such as -5. To them, numbers were closely bound up with *measurement*—of distance, area, and volume. Their algebraic rules were generally thought of in geometric terms, such as adding together various areas (see figure 4).

But if the Greeks had no use for negative numbers, they certainly did need fractions, or *rational numbers* as mathematicians call them. A (positive) *rational* number is one of the form a/b, where both a and b are natural numbers. Since b may be 1, the rational numbers include the natural numbers. (In the terminology of chapter 2, the natural numbers form a *subset* of the rational numbers.) Until some point in the sixth century B.C., the Greeks believed that the system of (positive) rational numbers was adequate for their geometrical purposes. But then they discovered to their horror that this was not the case. In particular, they found that the square root of 2 was not a rational number, which meant that rational numbers were not adequate for measuring, say, the hypotenuse of a right-angled triangle whose base and height both measure

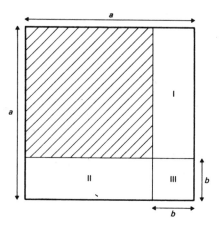

FIGURE 4   Greek algebra. The Greeks regarded familiar algebraic identities such as

$$(a - b)^2 = a^2 - 2ab + b^2$$

in purely geometric terms. To obtain the shaded area, that is, $(a - b)^2$, you start with the entire square (a^2), subtract the rectangle consisting of regions I and III (ab) and that consisting of regions II and III (also ab), and then add on the small square III (b^2) to compensate for the fact that this area was included in both subtracted rectangles. This produces the preceding identity.

1 unit (see figure 5). (To be able to measure all geometric lengths, the *real* numbers are required, about which more in a moment.) This discovery effectively marked the end of any progress in arithmetic made by the Greeks, who henceforth restricted their mathematics to geometrical construction.

Negative Numbers

The first systematic algebra to use zero and negative numbers was developed by Hindu mathematicians in the seventh century A.D. to handle financial transactions involving credit and debit. Besides being the first to use zero in a modern fashion, they wrote equations with negative numbers symbolized by a dot above the number (an early forerunner of

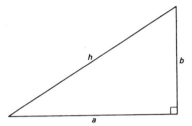

FIGURE 5   Pythagoras's theorem. For any right-angled triangle having sides a and b adjoining the right angle and h the hypotenuse, the identity

$$h^2 = a^2 + b^2$$

holds. When $a = b = 1$, this identity gives $h = \sqrt{2}$, an irrational quantity, which cannot be expressed as a quotient of two whole numbers.

our minus sign) and explicitly formulated a *law of signs* (plus times plus is plus, plus times minus is minus, minus times minus is plus). They also recognized that every positive number has two square roots—one positive and the other negative.

These early developments in India did not, however, affect the European mathematicians of the Renaissance period, from the fourteenth to the sixteenth century. Following the Greek tradition, they were happy to manipulate minus signs but did not recognize negative numbers as such. Indeed, they referred to negative roots to equations as "fictitious roots."

By the seventeenth century, some mathematicians were starting to use "negative numbers," but the tendency met with opposition, sometimes from prominent mathematicians. René Descartes spoke of negative roots as "false roots," and Blaise Pascal too thought that there could be no such thing as a number less than zero. Gottfried Leibniz, while agreeing that they could lead to absurdities, nevertheless defended negative numbers as a useful device in performing calculations. Leonhard Euler accepted negative numbers but believed they were greater than infinity (∞), reasoning that because $a/0 = \infty$, if we divide a by a number smaller than zero, the result must be greater than infinity.

It was during the eighteenth century that the algebraic use of negative numbers (denoted by the minus sign) finally became widespread, although even then many mathematicians still felt uneasy about them and would often go to great lengths to avoid using them if at all possible. Indeed, it is only when you adopt the axiomatic approach to numbers (see chapter 2) that negative numbers really "make sense." This conclusion applies equally well to complex numbers, but before we look at them, we should say something about "real" numbers.

Real Numbers

Although our discussion of number systems is divided according to the different types of numbers, historically there was no such distinction, and the theories of negative numbers, real numbers, and complex numbers all evolved over roughly the same period. Of the three, a proper treatment of the real numbers was by far the greatest accomplishment (being by far the most difficult), and although (as will be indicated) the theory of complex numbers assumes the prior existence of the real numbers, it was the theory of the real numbers that was the last to be worked out.

For all practical purposes, the rational numbers are more than sufficient. Indeed, in the real world (as opposed to the mathematical one), these are the only numbers used, with answers to problems being given to, at most, a few decimal places. Rational numbers also possess some pleasant mathematical properties. That is, if you add, subtract, multiply, or divide (except by zero) two rational numbers, the result is again a rational number. Moreover, the arithmetic of rational numbers satisfies all the axioms for an integral domain listed in chapter 2 (pp. 41–43). Mathematicians sum all this up by saying that the rational numbers constitute a *field*. A *field* is an integral domain in which division is possible, that is, a structure satisfying the seven axioms on page 41, together with this additional axiom:

8. For any number x other than 0, there is a number y such that xy = 1 (existence of multiplicative inverses).

It is an easy matter (try it) to verify that the y guaranteed to exist by axiom 8 is unique for any given x. Normally we write x^{-1}, or sometimes $1/x$, to denote this unique number y. Axiom 8 makes it possible to divide because a/b is the same as ab^{-1}.

Briefly then, a *field* is a structure that makes it possible to perform all the usual arithmetical operations with all the usual properties. Where the rational number field falls down is in its inability, as discovered by the Pythagoreans, to allow the solution of equations such as

$$x^2 = 2.$$

Using rational numbers, you can find a solution to any prescribed degree of accuracy:

$$1^2 = 1, \quad (1.4)^2 = 1.96, \quad (1.41)^2 = 1.9981,$$
$$(1.414)^2 = 1.999396,$$

and so on. But there is no rational number whose square is *exactly* equal to 2. The real numbers, on the other hand, constitute a field that includes the rational numbers and is rich enough to solve equations like the preceding one. The key idea in formulating this precisely is provided by the process of *successive approximation* inherent in the preceding example. The numbers 1, 1.4, 1.41, 1.414, . . . provide better and better approximations to a "number" whose square is 2. If it were possible to employ infinitely many places of decimals, we would be able to write down a number whose square was exactly equal to 2, namely,

$$1.414213 \ldots (ad\ infinitum).$$

But since we obviously cannot write down an infinite sequence of decimal places, how do we proceed in practice? By allowing the mathematics to handle the necessary infinite concepts for us, which means that the real numbers must be developed in an axiomatic way. This turns out to be an extremely difficult process, far outside the scope of a book like this. Indeed, when it was finally achieved during the 1870s, the formulation of a proper axiom system for the real numbers was one of the greatest accomplishments of mathematics.

The real numbers include all the rational numbers (just as the integers form a subset of the rational numbers), but many other numbers besides. A real number that is not rational is called an *irrational number.* Examples of irrational numbers are π, e, and $\sqrt{k}$ for any natural number k that is not a perfect square.

Complex Numbers

During the sixteenth century, European mathematicians—and in particular the Italian Rafaello Bombelli—began to realize that in the solution of algebraic problems, it was often useful to assume that negative numbers have square roots. It is perhaps understandable when we consider the climate of the time that such "numbers" were referred to as *imaginary numbers,* although to present-day mathematicians, all numbers are "imaginary" concepts, the square roots of negative quantities no more or no less so than any others. Nevertheless, it still is customary to refer to a square root of a negative real number as *imaginary,* thereby giving the word *imaginary* a special, technical, meaning in this context.

In fact, in order to have available the square roots of negative real numbers, it is necessary only to postulate the existence of a solution to the one equation

$$x^2 + 1 = 0.$$

If i denotes a solution to this equation (so $i^2 = -1$), then for any positive real number a, the square root of the negative number $-a$ will be $(\sqrt{a})i$. (Actually there are two square roots, $[\sqrt{a}]i$ and $-[\sqrt{a}]i$. Likewise, there are two solutions to the equation $x^2 + 1 = 0$, namely, i and $-i$.) It is the numbers of the form ai, where a is real, that are called *imaginary numbers.* The letter i was first used in this context by Euler.

A *complex number* is one of the form $a + bi$, where a and b are real numbers. The plus sign here is not intended to denote ordinary addition (how could it?); rather, it serves to separate the *real* part a of the complex number from its *imaginary* part bi. Notice that if $b = 0$, then $a + bi = a$, so the real numbers form a subset of the complex numbers. Similarly, if $a = 0$, then $a + bi = bi$, so the imaginary numbers also form a subset of the complex numbers.

At this stage, you may be thinking that calling something of the form $a + bi$ a number is not justified, even if you are prepared to accept $i = \sqrt{-1}$. But remember, it is not what numbers *are* that matters, but how they *behave.* If complex numbers have a workable and useful (either in mathematics itself or possibly in a wider context) arithmetic, possibly forming a field, they have as much right to be called "numbers" as do any others. So what is the arithmetic of complex numbers? The rules are given next. (For most people, this is the first number system that they actually meet from an axiomatic viewpoint. By the time that any of the integers, the rational numbers, or the real numbers are developed axiomatically, most people are already quite familiar with them.)

The rule for adding two complex numbers is quite simple: first you add their real parts together, and then you add their imaginary parts together. Thus,

$$(a + bi) + (c + di) = (a + c) + (b + d)i.$$

For example,

$$(2 + 3i) + (7 + 1i) = 9 + 4i,$$
$$(-3 + 4i) + (4 - 2i) = 1 + 2i.$$

Multiplying complex numbers is a little more complicated. The idea is to use the ordinary rules of algebra for multiplying two bracketed sums and then putting $i^2 = -1$. Thus,

$$(a + bi)(c + di) = ac + adi + bci + bdi^2$$
$$= ac + adi + bci - bd$$
$$= (ac - bd) + (ad + bc)i.$$

For example,

$$(2 + 3i)(5 + 7i) = 10 + 14i + 15i + 21i^2$$
$$= 10 + 14i + 15i - 21$$
$$= -11 + 29i$$

Perhaps surprising is that complex numbers can be divided. The rule is

$$\frac{a + bi}{c + di} = \frac{ac + bd}{c^2 + d^2} + \frac{bc - ad}{c^2 + d^2}i.$$

For example,

$$\frac{3 + 5i}{1 + 2i} = \frac{3 \times 1 + 5 \times 2}{1 + 4} + \frac{5 \times 1 - 3 \times 2}{1 + 4}i$$

$$= \frac{3 + 10}{5} + \frac{5 - 6}{5}i$$

$$= \frac{13}{5} - \frac{1}{5}i.$$

In fact, the complex numbers form a field. (As an exercise, you might like to check that the preceding definitions of addition and multiplication do lead to an arithmetic that satisfies all the field axioms.) So however strange you may feel the notion of a complex number to be, it does turn out to provide a "normal" type of arithmetic. In fact, it gives you a tremendous bonus not available with any of the other number systems. In the complex-number field, *every* polynomial equation can be solved! That is, if $a_0, a_1, \ldots, a_{n-1}, a_n$ are complex numbers, then there will be a complex number x which is the solution of the equation

$$a_n x^n + a_{n-1} x^{n-1} + \ldots + a_1 x + a_0 = 0.$$

This is not true of the real numbers, of course, as the equation $x^2 + 1 = 0$ demonstrates.

The result just mentioned is known as the *fundamental theorem of algebra*. It was first stated by Albert Girard in 1629 and then proved imperfectly by Jean-le-Rond d'Alembert in 1746 and Euler in 1749. The first totally correct proof was provided by Gauss in his 1799 doctoral thesis. So impressed was Gauss with the result that he subsequently offered three more, and distinct, proofs of it.

The fundamental theorem of algebra is just one of several reasons that the complex-number system is such a "nice" one. Another important reason is that the field of complex numbers supports the development of a powerful differential calculus, leading to the rich theory of functions of a complex variable. (This theory is touched on in chapter 8.)

Not only is the theory of complex numbers appealing mathematically, it is extremely useful as well. The first significant scientific use of complex numbers was made by Charles Steinmetz, who found them essential to performing efficient calculations concerning alternating currents. Indeed, nowadays no electrical engineer could get along without complex numbers, and neither could anyone working in aerodynamics or fluid dynamics. Einstein's theory of relativity uses complex numbers—the three spatial dimensions are regarded as real, and the time dimension as imaginary—and in quantum mechanics, physicists must deal with complex numbers.

Yet even though they constitute a field, even though they are very useful, and even though all number systems are purely abstract, "imaginary" constructs, many people still feel uneasy handling complex numbers. It is largely (entirely?) a matter of familiarity. For instance, real numbers may turn out to be extremely complicated mathematical objects when put under an analyst's "microscope," but there is always the comfortingly simple picture of the *real line* to fall back on—a "two-way infinite" straight line with 0 at the middle (see figure 6).

There is an equally comforting picture of complex numbers. Just as the real numbers can be thought of as the points on the real line, so the complex numbers can be identified with the points in a two-dimensional plane (see figure 7). The first person to propose this visualization of complex numbers was Caspar Wessel, a self-taught Norwegian surveyor who lectured on his idea in 1797. The same idea was rediscovered by Jean-Robert Argand, a Swiss bookkeeper, who published a book on the subject in 1806, and also by Gauss. Argand's book proved to have the greatest initial impact, and the *complex plane,* as the two-dimensional plane is properly called when intended as a representation of complex numbers, is sometimes still referred to as the *Argand diagram.*

Quaternions

Inspired by the representation of complex numbers as points on a plane, the Irish mathematician William Rowan Hamilton (1805–1865) developed an algebraic (i.e., essentially axiomatic) interpretation of complex

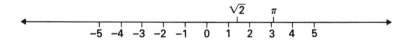

FIGURE 6   The real line. The axioms for the real numbers guarantee that the line is "continuous," with no "holes," not even infinitely small holes where a single point is left out (in the sense that the "rational line" has a "hole" where $\sqrt{2}$ ought to be).

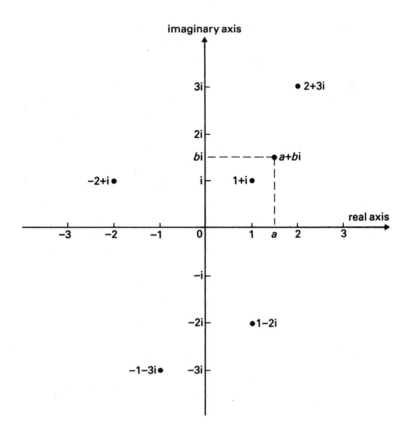

FIGURE 7   The complex plane. The complex number $a + bi$ is identified with the point having coordinates (a, b). The real numbers lie along the horizontal axis, and the pure imaginary numbers lie on the vertical axis.

numbers in terms of pairs of real numbers. He went on to investigate the possibility of a three-dimensional analogue of the complex number plane. This proved impossible, but as Hamilton discovered, if you work in four dimensions, it is possible to develop a system of what might be called *hypercomplex numbers.*

The *quaternions,* as Hamilton called his new numbers, did not come easily, and it was only after several years' thought that he was able to make the crucial breakthrough. As so often happens in mathematical research, the key insight did not come to him while he was seated at his desk. Instead, at dusk one day in 1843, he was strolling with his wife along the Royal Canal in Dublin when he realized that provided he dropped the commutative law for multiplication, everything else would work (i.e., he would get an otherwise acceptable number system). He was so excited about this realization that he stopped at Brougham Bridge to scratch the basic formulas on a stone. (The original graffito has long since weathered away, but the bridge now bears a plaque commemorating the great event.)

Briefly, a quaternion is a number of the form

$$a + bi + cj + dk$$

where a, b, c, d are real numbers, and i, j, k are "imaginary" numbers that satisfy the equation $i^2 = j^2 = k^2 = -1$. The key equations that Hamilton inscribed on the bridge are

$$ij = k, \quad jk = i, \quad ki = j,$$
$$ji = -k, \quad kj = -i, \quad ik = -j.$$

By using these rules, any two quaternions may be multiplied together (to give a third quaternion) by means of the ordinary rules of algebra. (Addition is simply term by term, as with complex numbers.) The resulting system of numbers satisfies all the axioms for an integral domain (pp. 41–43) except for the commutative law of multiplication.

Quaternions are used in modern physics, as is an even more bizarre number system—the *octonions*, an eight-dimensional number system in which both the commutative law of multiplication and the associative law are lost. But now it's time to return to the natural numbers and, in particular, to Gauss's work in number theory.

The Gaussian Integers

In 1796, Gauss proved a profound theorem of number theory called the *quadratic reciprocity law*, which concerns the solution of equations such as

$$x^2 \bmod 7 = 3,$$

which are of the form

$$x^2 \bmod p = q,$$

where p and q are prime numbers. In trying (with eventual success) to generalize his theorem to cover equations of higher orders ($x^3 \bmod p = q$, and so on) Gauss found that his calculations were made easier by working with numbers of the form $a + bi$, where a and b are integers (and $i = \sqrt{-1}$, as usual), rather than with integers alone. Nowadays such "complex integers" are known as *Gaussian integers.* They are particularly useful when factorization is involved. Just as the ordinary integers allow the factorization

$$a^2 - b^2 = (a + b)(a - b),$$

so the Gaussian integers give

$$a^2 + b^2 = (a + bi)(a - bi).$$

Intuitively, the Gaussian integers would appear to occupy the same position in the complex-number field as do the ordinary integers in the real-number field. But just how like the integers are the Gaussian integers?

As made clear in chapter 1, the most significant fact about the integers is encompassed by the *fundamental theorem of arithmetic,* that every integer can be expressed as a product of a unique collection of primes (numbers that cannot be split up any further) and possibly -1. Gauss demonstrated that among the Gaussian integers are certain numbers that are "prime" (i.e., cannot be split up) and that in terms of these "primes," an analogue of the fundamental theorem of arithmetic holds for the Gaussian integers: the *unique factorization theorem.* (The "primes" here are not numbers of the form $a + bi$, where both a and b are primes; rather, the Gaussian primes are defined as being those Gaussian integers that cannot be reduced to a product of other Gaussian integers. For this reason, mathematicians often refer to them as *irreducibles.*)

The Class Number Problem

The Gaussian integers turned out to be useful in other contexts besides reciprocity laws—notably their connection with Fermat's last theorem, about which I shall say more in chapter 10. So useful did they prove to be that it made sense to examine other, similar number systems, and this is what Gauss did. Among the various approaches that can be made, particularly fruitful are the systems of the form $a + b\sqrt{-d}$, where d is some positive integer other than 1.

At this stage, a minor surprise is in store. To obtain a reasonable number system (i.e., one that bears some resemblance to the ordinary integers), in the case where $d \bmod 4 = 3$, you must allow the numbers a and b to be half integers as well as integers. For example,

$$\tfrac{1}{2} + 2\sqrt{-3}, \quad \tfrac{3}{2} + \tfrac{5}{2}\sqrt{-3}$$

are numbers in the system that corresponds to $d = 3$. (If $d \bmod 4 \neq 3$ then, as with the Gaussian integers, a and b are just integers.)

Once you have made this minor modification, you can ask for which values of d you get a reasonable "number theory." In particular, for which values of d do you get a unique factorization theorem? For $d = 1$ (the Gaussian integers), $d = 2$, and $d = 3$, you do. But for $d = 5$, you do not. In this system, the number 6 (for example) has the two factorizations (into irreducibles)

$$6 = 2 \times 3, \quad 6 = (1 + \sqrt{-5}) \times (1 - \sqrt{-5}).$$

In Gauss's time, nine values of d were known for which the system of numbers $a + b\sqrt{-d}$ (for a and b varying as indicated) has a unique factorization theorem. They are

$$d = 1, 2, 3, 7, 11, 19, 43, 67, 163.$$

Are there any more values? Despite considerable efforts by Gauss and others, no one was able to find any. The next result was the discovery by Hans Arnold Heilbronn and Edward Hubert Linfoot in 1934 that there could be, at most, one extra value, and if it existed, it must be astronomically large. But was there a tenth value?

In 1952, one man knew that there was not. In that year Kurt Heegner, a retired Swiss scientist who did mathematics as a hobby, published what he claimed to be a proof that there was no tenth d, but no one believed him. (His paper was *very* hard to follow. Even so. . . .) The rest of the world had to wait another fifteen years before they knew the truth. In 1967, Harold Stark of the Massachusetts Institute of Technology and Alan Baker of the University of Cambridge independently (and using different methods) also proved that there was no tenth d, and this time the mathematical community was convinced. Motivated by their discovery, Stark and Baker started looking at Heegner's earlier work and, to their amazement, found that it was essentially correct. The neglected Heegner had been right after all!

And there you have the reason that that number 163 is so special, giving rise to those curious results mentioned at the beginning of this chapter. It is the largest value of d for which the number system $a + b\sqrt{-d}$ allows unique factorization. (Unfortunately, it is not possible to give any indication here of just how the unique factorization property for 163 relates to the earlier discussion of that number. That is strictly for the professional mathematician.)

Having disposed of those number systems $a + b\sqrt{-d}$ that do allow unique factorization, what can be said about the ones that do not? Once again, Gauss led the way. With each number system derived from some value of d, he associated a certain natural number $h(d)$ called the *class number* of that system. This class number gives a measure of the margin by which unique factorization fails. If the class number is 1 (which it is for each value of d in Gauss's list), then unique factorization holds. If $h(d) = 2$ (as it is for $d = 5, 6, 10, 13$, for example) then unique factorization just fails. For class number 3 (for $d = 23, 31, 59$), it fails a little more; for class number 4 (for $d = 14, 17, 21$), it fails still more; and so on. The bigger the class number is, the more ways there are of factoring numbers in the system into "primes" (of the system).

In article 303 of his *Disquisitiones Arithmeticae* (Gauss's monumental work, mentioned in chapter 1), Gauss described some extensive computations of class numbers and observed that for each class number k, there seemed to be a largest value of d for which $h(d) = k$. The largest d for which $h(d) = 1$ was (as far as he knew) $d = 163$, the largest d with $h(d) = 2$ seemed to be $d = 427$, and the largest d with $h(d) = 3$ was apparently 907. But Gauss was able neither to confirm that any of these values really was the largest nor to prove that there always was a largest d, although he conjectured that this was nevertheless the case.

The *class number problem* (which assumes the truth of Gauss's conjecture in order to make sense) is to determine for each class number k

the largest d for which $h(d) = k$. (Thus Heegner's 1952 result solved the class number problem for the case $h = 1$.)

Virtually no progress was made on the class number problem from Gauss's time until the twentieth century. In 1916, Erich Hecke proved that if a certain rather complicated statement known as the *generalized Riemann hypothesis* were true, then Gauss's conjecture that each class number corresponded to only finitely many values of d would follow. But since no one had (or indeed has) any idea of whether or not the generalized Riemann hypothesis is true, Hecke's result did not say very much—or at least not on its own, it didn't. But then in 1934, building on recent work by Max Deuring and by Lewis Mordell, Heilbronn proved Gauss's conjecture, based on the assumption that the generalized Riemann hypothesis was false. Since the hypothesis in question certainly must be true or false—even if we do not know which (or, bearing in mind the results described in chapter 2, cannot decide which)—Hecke's and Heilbronn's results taken together finally proved Gauss's conjecture.

Now that Gauss's conjecture had finally been established, the way was clear to solve the class number problem itself. But progress was painfully slow. First was Heegner's 1952 result for the case $h = 1$. Then, in 1967, as well as re-solving the case $h = 1$, Alan Baker and Harold Stark disposed of the case $h = 2$. But none of the methods that had been developed was able to handle any other cases.

The big breakthrough came in 1975, when a "partial" solution was obtained by Dorian Goldfeld of the University of Texas at Austin. By means of a long and difficult argument in analytic (complex) number theory, Goldfeld showed that if a certain (rather complex) mathematical object were available, the entire solution to the class number problem would follow. The object he required was a geometric curve of a certain shape* having some unusual special properties. Finding curves of the requisite shape was not the problem; rather, getting one with the

---

*In particular, the curve should have an equation of the form

$$y^2 = ax^3 + bx^2 + cx + d.$$

Such curves are called elliptic curves. They have a number of applications in number theory besides the one described here.

special properties was. Goldfeld tried hard to find one and failed, as did everyone else who looked at the problem.

In 1983, Zagier and Gross were successful in their search. Their key idea was to look for certain special points on the curve, and in honor of the long-neglected Heegner, they called these special points *Heegner points*. What the proof then amounted to was one enormous equation. Simply calculating the two sides of the equation took up 100 pages, and then Zagier and Gross had to pair off terms on each side to prove that the equation was correct. Despite its length, however, this part of the proof is what mathematicians refer to as "straightforward." What is remarkable is the fact that a single curve somehow controls the behavior of an infinite family of number systems.

After 183 years, Gauss's class number problem had finally been put to rest.

Suggested Further Reading

For a historical account of the development of number systems, see *Numbers: Their History and Meaning,* by Graham Flegg (André Deutsch, 1983).

Number systems are briefly described in *Numbers: Rational and Irrational,* by Ivan Niven (Random House, 1961). A more complete description can be found in *Foundations of Real Numbers,* by Claude Burrill (McGraw-Hill, 1967), and also in *The Structure of the Real Number System,* by Leon Cohen and Gertrude Ehrlich (Van Nostrand, 1963).

For a nice account of the kinds of number systems that figure in the class number problem, see chapter 8 of *An Introduction to Number Theory,* by Harold Stark (Markham, 1970). At a higher level, there is *Algebraic Number Theory,* by I. N. Stewart and D. O. Tall (Chapman & Hall, 1979).

A very sophisticated account of the final solution to the class number problem is provided by Don Zagier's article "*L*-Series of Elliptic Curves, the Birch-Swinnerton-Dyer Conjecture, and the Class Number Problem of Gauss," which appeared in the mathematical journal *Notices of the American Mathematical Society* 31 (November 1984):739–743. This article also lists other (high-level) sources that can be consulted by those who wish to do so. But I should stress that this work is very advanced and that even the majority of professional mathematicians would find the proof difficult to understand.

Finally, if you want to look at Gauss's *Disquisitiones Arithmeticae* yourself, an English version was published by Yale University Press in 1966. (The original was published in Leipzig in 1801.)

Beauty from Chaos

Beauty in Mathematics

Bertrand Russell wrote in his 1918 book *Mysticism and Logic* that "mathematics, rightly viewed, possesses not only truth, but supreme beauty—a beauty cold and austere, like that of sculpture."

Another famous British mathematician, G. H. Hardy, wrote in his book *A Mathematician's Apology* (1940):

> The mathematician's patterns, like the painter's or the poet's, must be *beautiful,* the ideas, like the colors or the words, must fit together in a harmonious way. Beauty is the first test; there is no permanent place in the world for ugly mathematics. . . . It may be very hard to *define* mathematical beauty, but that is just as true of beauty of any kind—we may not know quite what we mean by a beautiful poem, but that does not prevent us from recognising one when we read it.

Both writers here were thinking of a highly abstract form of beauty, an inner beauty that is known to all professional mathematicians but that, for most people, must remain forever unseen and very likely not even dreamed of. It is a beauty of logical form and structure, of elegance of proof, a beauty that can be glimpsed only after a long and arduous apprenticeship.

Or at least such was the case until the early 1980s, when the development of the electronic computer and, in particular, its graphics facilities gave rise to some new mathematical developments that changed everything. *Chaotic dynamics* is one of several names given to one new area of mathematics enabled by computers. Although some of the mathematics in this subject is as difficult and abstract as any other examples of the mathematician's art, the essential beauty of the resulting structures can be displayed on a computer screen for all to see, professional and layperson alike. Hard copies of computer-graphics displays formed the core of an exhibition organized by the German Goethe Institute, which began to tour the world in 1985, finding its way into both university mathematics departments and public art galleries. The film industry, too, was quick to appreciate the potential of the new mathematics, and increasingly, ideas from *complex dynamics* (another name for the same field) were used to create graphics for science fiction films.

Figure 8 shows one of many pictorial representations of the kinds of structures commonplace in this new field. (Many of them may be produced in color to highlight patterns not visible in a black-and-white representation.) Remarkable though it may seem, the complexity in figure 8 results from quite simple mathematics (though a detailed *analysis* can involve very advanced methods). That mathematics is explained in this chapter.

How Long Is the Coastline of Britain?

This was the question asked in an epoch-making article of the same title published in the magazine *Science* in 1967. The author was Benoit Mandelbrot, a brilliant French mathematician working at the IBM Thomas J. Watson Research Center at Yorktown Heights, New York. On the face of it, the question seems innocuous enough, and you might expect that a good answer could be found either with the aid of a map or by aerial reconnaissance. The only trouble is that no matter how carefully you try, you will not get the right answer, and for a very good reason: *there is no right answer!* Mandelbrot arrived at this startling conclusion by reasoning as follows:

Suppose you make your measurement by flying around the coastline in a jet at an altitude of 10,000 meters, taking photographs of the coast all the time, and then, using the appropriate scaling factor, calculating the total length as indicated by the vast collection of photographs you

FIGURE 8 Fractal art—a view into Mandelbrot's world.

will have accumulated. How accurate is this answer? Not very. From 10,000 meters, you will be unable to distinguish many small bays and promontories. (Let's assume that your camera is a good but otherwise ordinary model.) If you repeat the measurement from a small airplane flying at a height of 500 meters, so much extra detail will be visible that this answer will be significantly greater than your previous one. For example, what on the first photograph appeared as a smooth stretch of coastline will now be found to consist of numerous little inlets, bays, and promontories.

Now suppose you set off on foot to measure the coastline using a pair of dividers set at a separation of, say, 1 meter. Then features of the coastline not visible from the air will result in an answer that is even greater. If you repeat the measurement with the dividers set at 10 centimeters, the result will be bigger still. And so on. Each time you make your scale of measurement finer, more detail of the coastline will be visible, and your answer gets larger. Soon you will be measuring around pebbles, then around grains of sand, then molecules, and so on. And all the while your answer keeps on growing.

In the physical world, this process of taking finer and finer measurements must end eventually. Human limitations would probably bring you to a stop with the 1-meter dividers, whereas the physicist might argue that the procedure has a *theoretical* limit at the atomic level. But from the mathematician's idealized viewpoint, the process of making finer and finer measurements may be continued indefinitely. Since this means that the corresponding sequence of measurements will increase indefinitely, it follows that there is no *mathematically precise* notion of the length of the coastline, only arbitrary choices—choices that are not even approximations of some "real" answer.

An idealized, mathematical analogue of Mandelbrot's elusive coastline is provided by a geometrical figure first considered by Helge von Koch in 1904, which we call *Koch's island.* Figure 9(a) shows Koch's island as seen from a rocket in outer space. From this distance, it looks just like an equilateral triangle. But as the rocket approaches Earth, it becomes clear that each of the three straight edges actually contains a central, triangular promontory, forming an equilateral triangle occupying the middle third of the line (b). If the perimeter length in (a) is 3 units, then that in (b) will be 3 × (4/3) units. Coming in closer still, you see that each of the twelve straight edges you saw before likewise contains a promontory in the shape of an equilateral triangle occupying the middle third (c). The perimeter length now is 3 × (4/3) × (4/3) units.

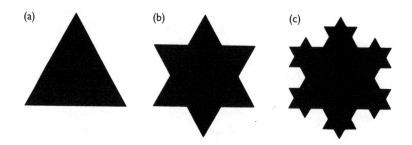

(a)   (b)   (c)

FIGURE 9   Constructing Koch's island.

FIGURE 10 Koch's island taking shape.

Figure 10 shows the island as seen from much closer, after several more levels of detail have unfolded, and gives some indication of the actual (?) shape of Koch's island.

To the mathematician, the nice feature of Koch's example is the regularity with which successive levels of detail appear. At each stage, the middle third of every straight-line segment of the coastline is replaced by two straight-line segments, each equal in length to that third, as shown in figure 11.

As you might surmise from examining figures 9 and 10, Koch's island does have a (mathematically) well-defined shape, which figure 10 approximates quite well as far as the human eye can discern. The mathematically precise *coastline* of Koch's island is the "curve" that is the limit of the infinite sequence of approximations to it, of which figure 9 gives the first three. At this point, the mathematics takes over from the

FIGURE 11   Generating the Koch coastline.

human cartographer. *Mathematically,* this limit curve is precisely determined and, like any other curve, consists of an infinitude of points
strung together to form a "line." The process of arriving at the limit
curve is analogous to arriving at the number 1/3 as the limit of the infinite sequence of decimals

$$0.3, 0.33, 0.333, 0.3333, 0.33333, \ldots .$$

Since Koch's island is a mathematically defined region of the plane,
it has a definite area. The actual numerical value of its area will depend
on the units of measurement being used, of course, but it will certainly
be *finite.* (It may be calculated as a limit of a sequence of numbers, much
like the preceding 1/3 example; it is, in fact, exactly 1.6 times the area
of the triangle in figure 9[a].) What about the length of the coastline
surrounding this finite area? Each successive stage of the Koch process
increases the length of the "coastline" by a factor of 4/3. By the time the
Koch curve (as the limiting coastline is called) is reached, this 4/3 increase will have occurred infinitely often, and so the length of the Koch
curve will be infinite.

How can a finite area have an infinite boundary? Figures 9 and 10
themselves provide the answer. The boundary curve twists from side to
side along its entire length. For each of the finite approximations to the
final curve, you can draw this twisting in full if you use a suitable scale
(magnification), but for the actual Koch curve, the twisting is infinite,
and then something very strange occurs: you enter a new dimension.

New Dimensions

The curves that we usually meet in geometry all are *one dimensional,*
that is, a creature living on, say, a straight line or a circle can travel in

only one direction (if traveling backward is regarded simply as negative forward movement). The usual geometrical surfaces such as planes or spheres are *two dimensional*: a creature there has two independent directions of travel, often referred to in terms of forward/backward and left/right. Solid objects are *three dimensional*, allowing for three directions of motion. Railways are an example of motion restricted to one dimension; ships can travel in two dimensions over the surface of the sea; and aircraft can move in three dimensions.

As far as human experience is concerned, the universe we live in has only three dimensions (although relativity theory regards time as a "fourth dimension" and some current physical theories ascribe to the universe eleven dimensions, the three we are physically aware of plus another eight that manifest themselves as the basic forces of nature, gravity, magnetism, and so on). For mathematicians, however, there is nothing special about three "dimensions." "Spaces" of four or more dimensions may, and routinely are, considered. Although they cannot be realized by traditional geometry, such higher-dimensional spaces can have real, practical use. (A case in point is the subject known as *linear programming*, discussed in chapter 11.) But notice that such "higher dimensions" are still whole numbers.

Where does the Koch coastline fit into all this? Being a curve (in the mathematical sense, although with infinite twisting you could not hope to draw it), you might imagine that it is one dimensional, but this is not so. Although each of the approximations to the Koch curve obtained by means of the process described earlier is one dimensional, the limiting curve is not. With the direction changing infinitely often, we are no longer in a familiar world, and indeed our use of the word *direction* can no longer be justified. So we cannot hope to decide on the dimensionality of the Koch curve by speaking of the "direction of travel." Instead, what we must do is find another way of getting at the concept of dimension, one that does not depend on direction.

It makes sense to adopt an approach suited to the nature of the Koch curve. The key feature is *self-similarity*: the parts are similar to the whole, only on a reduced scale.

Suppose we take a D-dimensional figure and divide it into N entirely similar parts. Then the *similarity ratio, r,* between the entire figure and a single part (i.e., the factor by which the whole exceeds the part in size) is given by

$$r = \sqrt[D]{N}.$$

(Since the figure is D dimensional and r must be evaluated "along a dimension," it is necessary to take the Dth root of N.)

For example, suppose we split a straight line into N equal pieces (see figure 12). Then each piece is exactly $1/N$ of the length of the whole, so the similarity ratio will be N. This is precisely the value obtained from the preceding formula when you make $D = 1$.

Or how about splitting a rectangle (so $D = 2$) into N pieces by dividing it horizontally and vertically into k segments (see figure 13)? Then the entire rectangle is split into exactly $N = k^2$ identical smaller replications of the whole, and the (linear) ratio r of the whole to any one of the parts is given by

$$r = \sqrt[D]{N} = \sqrt[2]{N} = \sqrt[2]{k^2} = k.$$

Again, this is exactly what you would expect.

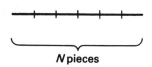

N pieces

FIGURE 12   Self-similarity for a straight line.

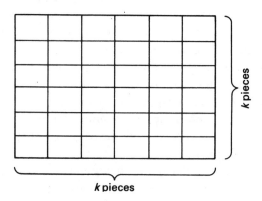

k pieces

FIGURE 13   Self-similarity for a rectangle.

In both these cases, we seem to have been going around in circles, but this was because we were dealing with cases that are very familiar and not problematical. When we apply the same analysis to the Koch curve, we arrive at a more surprising conclusion. For this curve we do not know D, but the values of N and r are easily determined. All we need to do is look at the replication procedure that produces the curve. We first look at one section of coastline (see figure 11[a])—any one will do, since they all are the same. In replication (see figure 11[b]), the single line is replaced by four lines (so $N = 4$), each one-third the length of the original line (so $r = 3$). Since this is true for any one piece of the coastline, it will be true for the whole Koch curve. So, according to the formula just determined,

$$3 = \sqrt[D]{4}.$$

What is D? Certainly not a whole number. The only way to determine its value is by using logarithms. If you take logarithms of both sides of the preceding equation, you get

$$D \log 3 = \log 4.$$

By referring to a book of logarithm tables (or using a calculator that can evaluate logarithms), you may calculate D to four decimal places; it is

$$D = 1.2618.$$

Thus the Koch curve is a mathematical entity whose dimension is *fractional.*

It is not just "curves" that can have fractional dimensions. Equally bizarre "surfaces" and "solids" also may be constructed using self-replication processes. For instance, by starting with a cube and successively removing "middles," you will eventually arrive (i.e., after infinitely many repetitions) at an object known as the *Sierpinski sponge* ($D = 2.7268$), whose construction is shown in figure 14. This incredible object has zero volume enclosed by an infinite "surface." Each external face of the sponge is known as a *Sierpinski carpet* (after the mathematician Wacław Sierpinski) and has zero area surrounded by an infinite boundary. The dimension of the Sierpinski carpet is $D = 1.8928$ ($N = 8$, $r = 3$). You should

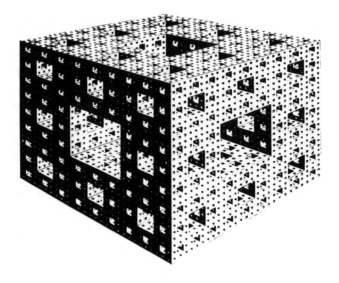

FIGURE 14   The Sierpinski sponge taking shape.

be able to confirm both values of D associated with the sponge by looking at figure 14 and using the formula

$$r = \sqrt[D]{N},$$

or, taking logarithms,

$$D = \log N/\log r.$$

Figures having a fractional dimension were given the name *fractals*
by Mandelbrot in 1977. *Fractal geometry* is the study of such objects.

The remainder of this chapter also is concerned with fractals, but
ones of a different sort from the Koch curve and the Sierpinski sponge.
These two fractals are highly regular. The self-replication process is the
same at every level, and "zooming in" on a particular part of the figure
to see more detail produces no surprises—just more of the same, ad infinitum. Since 1980, computers have been used to examine fractals
whose self-replication is continuously changing (although, as will become clear, it often can still be called self-replication). Zooming in on

such figures can produce totally unexpected results, of which figure 8 is an example. The examination of such fractals is a subject that is part mathematics and part experimentation (with the computer as the experimental tool), and it leads the researcher into a fascinating and often extremely beautiful new world. Like many other new worlds, its discovery was governed partly by chance.

Discovering a New World

By 1978, Benoit Mandelbrot's work on fractals was very well developed. The previous year had seen the publication of his book *Fractals: Form, Chance, and Dimension,* which showed how many everyday phenomena in physics, biology, and mathematics give rise to fractals. All the fractals he considered were, like the Koch curve, self-similar. They produced some interesting mathematics and occasionally startling conclusions, as well as some attractive and highly symmetrical figures (many of which are illustrated in Mandelbrot's book), but his examples had a built-in predictability that is not present in the real-life fractals of which they were mathematical counterparts. (For instance, Britain's coastline exhibits a fractal behavior far less ordered than Koch's coastline.) This extreme orderliness and predictability arises because the fractals considered are self-similar in the sense of scaling and translation (in mathematical language, they are *invariant under linear transformations*). Working with Mark Laff at IBM in 1978/79, Mandelbrot began to investigate fractals that are invariant under *nonlinear transformations* (where, instead of a simple scaling, more complicated manipulations are allowed, involving squaring, cubing, and so on). In such a case, the only way to get any idea of what the corresponding fractal looks like is to have a computer start generating it. Indeed, during the early part of this century, work on the same notions by Gaston Julia and Pierre Fatou in France came to a halt partly because there was no way of picturing the objects being considered. (Mandelbrot had been aware of this work from his student days at the École Polytechnique in Paris, where Julia had been one of his teachers.)

By the end of 1979, Mandelbrot had come to the conclusion that it was worth investigating—using the computer—the behavior of the particular function $x^2 + c$, where both the variable x and the constant parameter c are complex numbers. (Precisely what "behavior" was being considered will be explained later, but suffice it to say now that it is possible to use computers to draw diagrams relating this behavior to varying values of the parameter c.)

Ironically enough, Mandelbrot was not at IBM for what was to be the
crucial year, 1979/80, but was visiting Harvard University and so did
not have daily access to the famed "unlimited computing facilities" of
IBM at the time when his work most needed them. But in the basement
of the Science Center at Harvard, he found a newly delivered VAX superminicomputer, to which were attached a rather old Tektronix visual
display unit to view the output and a Versatec printer that could provide
hard copies. A Harvard teaching assistant named Peter Moldave volunteered his services as an (unpaid) programmer for the project. And so
the work went ahead.

The first picture they obtained was a crude version of the beetlelike
double blob shown (in much greater detail) in figure 20 (on page 105).
This they had been expecting, as the theory had predicted it. More perplexing were a number of smaller blobs away from the main picture.
Closer examination of these blobs revealed that they were smaller versions of the main beetle! It looked as though the familiar fractal reproductive behavior was once more manifesting itself. By performing more
accurate computations, Mandelbrot and Moldave obtained better pictures showing more detail, until suddenly the pictures began to look increasingly messy. Perhaps their ancient graphics equipment was faulty?
To make sure, Mandelbrot ran the program again on an IBM mainframe
computer at his home base at Yorktown Heights. Not only did the mess
fail to disappear, but a better-quality picture showed that it had an underlying pattern. Zooming in to take an even closer look, Mandelbrot
and Moldave found that some of the small dustlike blobs were not
smaller versions of the beetle, as they had supposed, but instead were
beautiful and intricate patterns—spirals, families of figures like sea
horses, and other fascinating shapes (see figures 8 and 23). Mandelbrot
had glimpsed his new world.

Order and Chaos

Order and chaos. Throughout history and throughout the universe, the
two have competed for supremacy. Often only a knife edge separates

them: a small change in the pressure can turn the ordered flow of water from a tap into a highly complex chaos of vortices; ordered animal populations (including human ones) can be transformed with frightening ease into uncontrollable anarchies. In the other direction, order can emerge from chaos, as witnessed by the evolution of life—and ultimately humankind—from the formal chaos of the universe. As we shall see, the transition from order to chaos and the subsequent emergence of order from within that chaos appears in dramatic form with the study of simple *feedback loops*.

The essential feature of the feedback mechanism is that a quantity, call it x, changes, either over time (as in the following example) or else with respect to another variable, in such a way that the value of x at any instant regularly depends on its value at the previous instant (see figure 15). Processes of this kind permeate all the exact sciences and most, if not all, of the inexact sciences. In fact, much of modern mathematics was developed to deal with such processes. For example, it was to handle the case in which the increment between the old x and the new x is infinitesimal that the many techniques for handling differential equations were developed.

To study a feedback process mathematically, the rule for generating the new value of x from the previous value is taken to be a mathematical function, $f(x)$. Then, starting with an initial *seed* value, x_0, of x, successive values $x_1, x_2, x_3, \ldots$ are generated according to the rule illustrated in figure 16. This can be done for any function $f(x)$, but the resulting feedback process will not be very interesting unless the chosen $f(x)$ is something other than a *linear* function, one having the form

$$f(x) = ax + b$$

for some constants a and b. We shall be particularly concerned with the case in which $f(x)$ involves a parameter, as the choice of that parameter

FIGURE 15 A feedback mechanism changing the value of x.

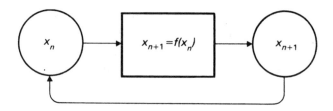

FIGURE 16   Generating successive values of *x* by feedback.

can have a dramatic effect on the behavior of the resulting feedback process.

It is customary to regard the operation of a feedback loop as a *dynamical system,* which sends an initial point x_0 successively into the points $x_1, x_2, x_3, \ldots$. The sequence of points to which x_0 is sent may be called the *path* or *orbit* of x_0. If this path is ordered, we may speak of *ordered dynamics;* if it is not, it may be described as *chaotic dynamics.* This nomenclature alone should indicate how this study relates to many everyday phenomena.

As an example, consider the growth of a population over a number of years. Suppose the initial size of the population is x_0, and let x_n be the population after *n* years. The growth rate during the $(n + 1)$th year is then

$$r = \frac{x_{n+1} - x_n}{x_n}.$$

If the growth rate is constant from year to year, this equation will be valid for every value of *n* and may be rearranged to give the (linear) *dynamical law*

$$x_{n+1} = f(x_n) = (1 + r)x_n.$$

After *n* years, the population will be

$$x_n = (1 + r)^n x_0$$

(an expression obtained by working backward from $x_n = (1 + r)x_{n-1}$, $x_{n-1} = (1 + r)x_{n-2}$, and so on, down to $x_1 = (1 + r)x_0$). This is an example of what is known as *exponential growth* and is typical (up to a point) of many real-life phenomena besides population growth. As should be clear from what we found in chapter 1, if continued without check for a number of years, such a growth mechanism will lead to enormous populations. What happens in practice is that such growth occurs over only a certain period, after which a limit is reached. In 1845, Ferdinand Verhulst formulated a growth law that takes account of the existence of a maximum possible population size, say X. Verhulst's law says that the growth rate will drop from r to 0 as the population approaches X. A simple way to represent this mathematically is to replace the constant growth rate r by the variable growth rate $r - cx_n$, where c is some constant. Since the population growth should become zero when $x_n = X$, the value of the constant c will have to be r/X. With this value, the dynamical law for Verhulst's process is

$$x_{n+1} = f(x_n) = (1 + r - cx_n)x_n = (1 + r)x_n - cx_n^2.$$

Once the value X has been reached, the population remains constant:

$$f(X) = X.$$

If the population is smaller than X, it will grow larger, and if it is larger than X, it will grow smaller. If you try it out (either by hand or using a computer) you will see that the Verhulst process gives a population evolution that eventually becomes stable at size X, regardless of the initial size. At least, that is what happens if r is less than 2 (i.e., a 200 percent growth rate)—a restriction that certainly applies to the growth of human populations. But as the meteorologist Edward N. Lorenz observed in 1963, for larger values of r, the Verhulst law describes certain aspects of turbulent flow, and it has other applications in laser physics, hydrodynamics, and the theory of chemical reactions, so the behavior of the Verhulst process for values of r greater than 2 is not without interest. And it turns out that this is where the really fascinating results are found.

Making $c = r/X$, the preceding rule becomes

$$x_{n+1} = (1 + r)x_n - (r/X)x_n^2.$$

By appropriately altering the units of measurement, we may assume that $X = 1$, so the rule is simplified even further to

$$x_{n+1} = (1 + r)x_n - rx_n^2 = x_n + rx_n(1 - x_n).$$

Using a typical home computer, anyone can easily perform some experiments to see how the Verhulst process evolves for different values of r in this last formulation of the equation, starting with the seed value (say) $x_0 = 0.1$ in each case. (To do this, the program should read in the chosen value of r, set $x = 0.1$, iterate the operation

$$x = x + r * x * (1 - x)$$

500 times to give the process time to settle down, and then calculate and print out the next twenty or so values of x.) For values of r less than 2, the process quickly settles down to the equilibrium value of $x = 1$. For r just above 2, the process settles into a regular oscillation between two values ($r = 2.1$ gives the values 0.82 and 1.13). This behavior continues for all choices of r up to $r = 2.5$, when there is a repeated cycling through four points (0.54, 1.16, 0.70, 1.23). This continues up to $r = 2.55$, when a cycle through eight values begins. For $r = 2.565$, it doubles up yet again to sixteen values through which the process eventually cycles indefinitely, and so on and so on, continuing to double ever more frequently until at $r = 2.57$ the doubling-up effect has occurred infinitely often. At this point, the behavior of the dynamical system becomes chaotic, jumping around all over the place with no apparent pattern.

The various cycles to which the process converges for values of r less than 2.57 are known as *attractors*. So for r less than 2, the attractor consists of one point, the fixed point $x = 1$; for r between 2 and 2.5, the attractor is an oscillating pair of values: for r between 2.5 and 2.55, it is a cycle of four points; and so on.

The picture of what is going on can be made clearer by drawing a graph to relate the behavior of the process (after the initial settling-down period) to the appropriate value of r. The main part of figure 17 shows the result of taking values of r from 1.9 to 3.0, measured along the horizontal axis, and plotting 120 successive values of x after an initial settling-down period of 5000 iterations.

A careful analysis of the chaotic region above $r = 2.57$ reveals that beneath this chaos lies a considerable amount of order. For instance, near $r = 3.0$, there is only one chaotic band; at $r = 2.679$, this splits into two chaotic bands; at $r = 2.593$, into four; then into eight, sixteen,

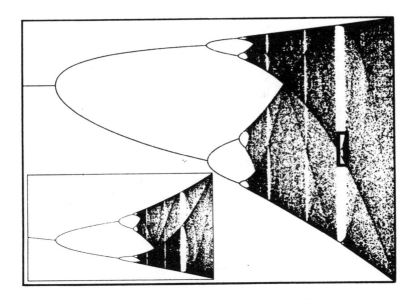

FIGURE 17   The Verhulst process $(1.9 < r < 3.0)$, with a blowup of the area indicated, showing self-replication. Values of r from 1.9 to 3.0 are plotted along the horizontal axis. For each value of r, 120 successive values of x are plotted against the vertical axis after an initial run of 5000 iterations to allow the process to settle down. For values of r less than 2, only one x-value arises. For r between 2 and 2.5, there are two values; for r between 2.5 and 2.55, four; then eight values as far as 2.565. This doubling up continues ever more rapidly until $r = 2.57$, where chaos sets in. But within this chaos, order begins to reemerge, including self-replication.

and so on, doubling up each time until at $r = 2.57$, this doubling up has occurred infinitely many times, with the entire process replicating the behavior of the dynamical system itself. In fact, there is a *universal constant* related to this doubling-up process. It is called the *Feigenbaum number* (after Mitchell Feigenbaum, who first discovered the number), and its value to ten decimal places is

$$f = 4.6692016609.$$

The range of values of r for which you get two chaotic bands is f times wider than the range for which you get four bands; the range of values of r for which you get four chaotic bands is f times wider than the range for which you get eight bands; and so forth.

The constant f arises in many other examples of feedback systems in which ordered behavior changes into chaos.

A particularly intriguing feature of the behavior depicted in figure 17 is the appearance of bands in the chaotic region, where order seems to reign (briefly) once more. For instance, near $r = 2.83$, the chaos suddenly gives way to a three-cycle attractor. And in the region around its middle point, you will discover a tiny replica of the entire Verhulst diagram, complete with its own ordered bands amid the chaos. (An enlargement of this region is shown in the inset of figure 17, "stretched" in the horizontal direction.) Fractal behavior once more! But this is only the beginning.

Julia Sets

Mandelbrot's work in 1980, mentioned earlier, was a continuation of work (mainly by P. J. Myrberg in the 1960s) on the Verhulst process. The main difference in Mandelbrot's approach was to allow the variable and the constant parameter to be complex numbers rather than just real numbers. So instead of sending numbers to numbers (on the real line), it sends points to points (in the two-dimensional complex plane, or Argand diagram).

To make things slightly easier, instead of taking the function

$$f(x) = x + rx(1 - x) = rx^2 + (1 + r)x,$$

as before, Mandelbrot used the slightly simpler formula

$$f(x) = x^2 + c,$$

which is the function we shall consider here.

Suppose you start with some seed value x_0 (a complex number) and then see what happens when you iterate the function f to generate a sequence of points $x_0, x_1, x_2, \ldots$ according to the rule

$$x_{n+1} = f(x_n).$$

The results obtained for the Verhulst process suggest that the choice of the constant parameter c will have a significant effect. Suppose we look at the simplest case, when $c = 0$. The dynamical law is then just

$$x_{n+1} = x_n^2.$$

There are three possible outcomes, depending on the choice of x_0. First, if x_0 is less than a distance 1 unit from 0 (the origin), the numbers in the sequence will become smaller and smaller (i.e., closer and closer to 0), which is to say that 0 is an attractor for the system. Second, if x_0 is a greater distance than 1 unit from 0, the numbers in the sequence will get larger and larger, in which case we say that infinity is an attractor (though because infinity is not a *point* in the complex plane, this particular use of the word *attractor* is a matter of convention). The remaining possibility is when x_0 is exactly 1 unit from the origin (i.e., x_0 lies on the unit circle centered at 0). In this case, the sequence never leaves the unit circle. Thus the unit circle is the *boundary* between the two domains of attraction, one governed by 0 and the other by infinity.

Insofar as it divides the complex plane into two distinct regions of attraction separated by a bounding curve, this example is typical of all the cases considered by Mandelbrot (and others). But Mandelbrot discovered that for nonzero values of the parameter c, not only can the noninfinite attractor consist of more than one point, but also the boundary between the domains of the two attractors can be incredibly complex and extremely beautiful.

For $c = 0.31 + 0.04i$, for example, the noninfinite attractor is a single point, but the boundary between its domain and that governed by infinity is not a perfect circle; rather, it is the attractively deformed circle shown in figure 18(a). It is a fractal deformation: if you move in to take a closer look at any one part of the boundary (using a computer as your "microscope"), you will find the familiar, endlessly repeating self-similarity that is characteristic of fractal curves.

Although it was only with the advent of the computer that it became possible to examine such figures, Julia and Fatou had already proved that any piece of the boundary, no matter how small, will contain all the information required to determine the whole curve—in that the entire boundary can be generated by repeatedly subjecting the piece to the transformation that determines the system, in this case $f(x) = x^2 + c$. In honor of Julia, such boundary sets are nowadays known as *Julia sets*.

Figure 18(b) shows a Julia set associated with a dynamical process having a noninfinite attractor consisting of a cycle through three points. The dynamical law here is $f(x) = x^2 + c$, with $c = -0.12 + 0.74i$. Figure 19 shows other examples of Julia sets that arise from the law $f(x) =$

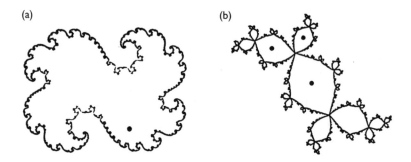

FIGURE 18   Julia sets with their attractors (see the text for details).

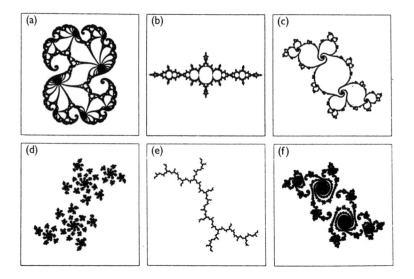

FIGURE 19   Julia sets from the boundary of the Mandelbrot set (see the text for details).

$x^2 + c$, including some extreme cases in which the regions degenerate into "dust" or "dendrites" (the details are discussed later).

The diversity of structure exhibited by the Julia sets for differing choices of the parameter shows just how critical the choice of c is. A natural question to ask is whether the values of c have any discernible pattern for which the corresponding dynamical system and its Julia sets have a particular form. Investigation of this question led Mandelbrot to his 1980 discovery of the region (subset) of the complex plane that now bears his name: the Mandelbrot set.

The Mandelbrot Set

The black beetle-shaped blob shown in figure 20 is known as the *Mandelbrot set*. Since its discovery, it has been shown that this set is intimately connected with the behavior of all dynamical processes, not just the single example considered here. As such, it occupies a special, fundamental place in mathematics, along with other special shapes such as the circle and the regular polygons.

As should be clear from a glance at figures 18 and 19, a complex dynamical process either divides the complex plane into one or more interior regions and a single exterior region stretching to infinity (figure 18[a] and [b], figure 19[a], [b], [c]), or it causes the Julia set to degenerate into a set that does not border an inner region (figure 19[d], [e], [f]). The exact behavior depends on the location of the parameter c rel-

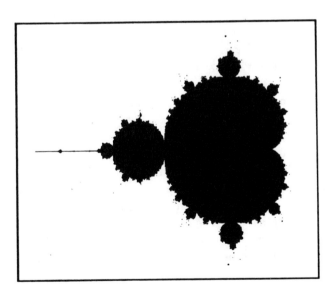

FIGURE 20   The Mandelbrot set ($-2.25 <$ Re $c < 0.75$, $-1.5 <$ Im $c <$ 1.5) (see the text for details).

ative to the Mandelbrot set. Continuing with the example of the process $f(x) = x^2 + c$, we first consider those cases with a nondegenerate Julia set. In this case, there is an attractor other than infinity.

If c is chosen from the main body of the Mandelbrot set, the associated dynamical system will have a noninfinite attractor consisting of a single point—a fixed point, satisfying $f(x) = x$. The Julia set in this case is a fractally deformed circle, as in figure 18(a). (The constant c here is located near the right-hand edge of the cardioid-shaped main body of the Mandelbrot set.)

If, on the other hand, c is chosen from one of the buds attached to the main body of the Mandelbrot set, the Julia set will consist of infinitely many fractally deformed circles surrounding the points of a periodic-cycle attractor, and the points that are eventually sent to them. For figure 18(b), for instance, c is chosen from the center of the large bulb at the top of the Mandelbrot set. The three points indicated form the 3-cycle of the noninfinite attractor for the system. A point chosen from any of the three regions containing this attractor will move directly toward that 3-cycle; points from the other regions will be attracted to a local "attractor" that is sent into the 3-cycle.

If c is the germination point of a bud on the Mandelbrot set, the Julia set will have tendrils that reach toward a marginally stable attractor, as in figure 19(a), which eventually settles into a 20-cycle ($c = 0.273\ 34 + 0.007\ 42i$), or figure 19(b), which has a 4-cycle ($c = 1.25$).

Finally, if c is any other boundary point of the Mandelbrot set, the Julia set will be what is known as a *Siegel disk* (after Ralph Siegel), an example of which is shown in figure 21 ($c = -0.390\ 54 - 0.586\ 79i$). Here there is a fixed point surrounded by *invariant circles*. In this case, a point within the region bounded by the Julia set gradually makes its way toward the disk containing the fixed point, whereupon it will orbit forever around the fixed point on its invariant circle.

The four types of Julia set just described are the only ones possible for the process $f(x) = x^2 + c$. (In 1983, Dennis Sullivan showed that there is one other type of nondegenerate Julia set that may arise in other complex dynamical systems—a Herman ring [named after Michael Herman].)

So much for the nondegenerate Julia sets. But what about the others, as in figure 19 (d), (e), and (f)? Large-scale pictures of the Mandel-

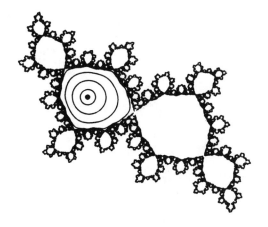

FIGURE 21 A Siegel disk (see the text for details).

brot set show that it is surrounded by hair-like, branching antennae. If c is chosen from one of these antennae, a similar-shaped Julia set will be obtained. Figure 19(e) shows the example for $c = i$. The behavior here is that the only attractor is infinity and that all points are sent there except for the ones actually on the hairlike Julia set.

Figure 20 is not sufficiently detailed to show the antennae themselves, but the location of some of them can be discerned from some isolated dots that lie on their path. Dots? A close examination (by a computer blowup) would reveal that they are nothing but tiny replicas of the Mandelbrot set itself! They even have tiny antennae around them, along which may be found . . . and so on, ad infinitum. (The gaps in the chaotic region of the Verhulst process (see figure 17) correspond to the positions of these "sprouts" relative to the real axis.) If c is chosen from one of these sprouts, the corresponding Julia set will be a combination of a dendrite and infinitely many copies of the Julia set from the corresponding value of c in the main Mandelbrot set (see figure 22).

The only remaining possibility is to choose c from outside the Mandelbrot set (with all its attachments) altogether. Here infinity is the only attractor, and the Julia set dissolves into isolated points called *Fatou dust*. This dust gets thinner and thinner the farther c is from the Mandelbrot set. If c is chosen from a point near the boundary of the Mandelbrot set,

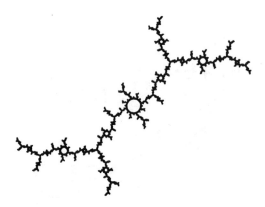

FIGURE 22   A Julia set from an outlining Mandelbrot sprout (see the text for details).

the dust is sufficiently thick to create fascinating patterns, as in figure 19(d) and (f). (For figure 19[f], c is close to the value that produces figure 19[c], and there is a noticeable similarity between the two Julia sets.) Such patterns in the dust are always fractal-like (i.e., self-similar), with chaotic dynamics.

Not surprisingly, with the boundary of the Mandelbrot set playing such a critical role in the dynamics of the associated systems, this boundary is itself of great interest. As you are probably expecting by now, this boundary region also has a complicated, fractal surface. Figure 23 provides just one glimpse into this incredible world, a world in which every single point of territory is contested, a world that is accessible only via the computer—with the degree of detail that can be discerned depending on the power of the computer. If any branch of mathematics is truly of this computer age, surely this must be it.

Suggested Further Reading

The world just described can be truly appreciated only with the aid of color representations, in which colors can be used to provide a kind of relief or contour map reflecting the system's dynamics. An excellent compilation of both color and black-and-white pictures (with explana-

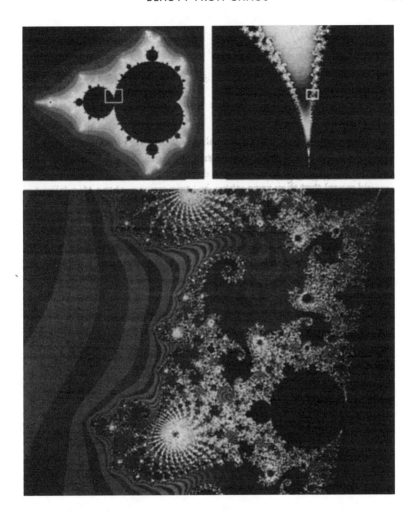

FIGURE 23   A journey into the border region of the Mandelbrot set.

tions) can be found in *The Beauty of Fractals,* by H. O. Peitgen and P. H. Richter (Springer-Verlag, 1986), a book that I recommend highly.

For a highly readable account of fractal geometry and its many applications, see *The Fractal Geometry of Nature,* by Benoit B. Mandelbrot (Freeman, 1982).

Simple Groups

The Enormous Theorem

Some time during the summer of 1980, Ohio State University mathematician Ronald Solomon put down his pen after solving a technical problem in algebra and, with that one simple action, ended a quest that had begun in the 1940s and involved more than a hundred mathematicians from the United States, Britain, Germany, Australia, Canada, and Japan. What Solomon had done was to fill in the last piece of an enormous and highly complex puzzle: the classification of the finite simple groups.*

The *classification theorem* is by far mathematics' biggest theorem. The initial proof ran to nearly 15,000 pages spread across some 500 articles in mathematical journals, and more than 100 mathematicians contributed to that proof. Along the way, discoveries were made that led to advances in the theory of computer algorithms, in mathematical logic, in geometry, and in number theory, and some people have speculated that the classification theorem may even have applications in the formulation of a unified field theory in physics.

And yet, as with other deep results in mathematics, the story of the

---

*All technical terms will be explained in due course.

classification theorem has very humble beginnings: in this case the familiar formula

$$x = \frac{-b \pm \sqrt{b^2 - 4ac}}{2a}$$

for the roots of the quadratic equation

$$ax^2 + bx + c = 0,$$

and attempts to obtain similar solutions for higher-degree equations (i.e., involving powers of x greater than 2, such as the cubic equation— see momentarily). Here, a "similar solution" means one that uses only the basic algebraic operations of addition, subtraction, multiplication, and division, as well as the extraction of roots. Such solutions are sometimes referred to as "solutions by radicals."

Examination of ancient tablets makes it clear that the Babylonian mathematicians of 1600 B.C. knew how to solve quadratic equations, although they had no algebraic notation to express their equations and solutions as we do nowadays. The solution (by radicals) of a cubic equation, that is, one of the form

$$ax^3 + bx^2 + cx + d = 0,$$

was not discovered until the sixteenth century, when the Italian mathematicians Scipione del Ferro and Niccolo Fontana independently devised the method. Girolamo Cardano published Fontana's solution in his book *Ars Magna* in 1545, and this volume also contained Ludovico Ferrari's method for solving a quartic equation (by reducing it to a cubic). But there the matter seemed to end. Despite the efforts of many mathematicians—including Euler in the middle of the eighteenth century—no one was able to obtain a solution to the quintic equation

$$ax^5 + bx^4 + cx^3 + dx^2 + ex + f = 0.$$

In 1770, Joseph Louis Lagrange suggested that such a solution (i.e., one by radicals) might not be possible, and in 1824, the Norwegian mathematician Niels Henrik Abel proved that this was indeed the case.

If there is no general method (i.e., no formula) for solving a quintic equation, it is natural to ask whether there is any way of deciding whether or not a *given* quintic equation can be solved (by radicals). Abel himself was wrestling with this problem when he died in 1829 at the age of 26. By this time, the young man who was eventually to solve the problem was also working hard on it. But the remarkable results obtained by the young Évariste Galois were not recognized by the mathematical community until eleven years after his death in a duel. And thereby hangs one of mathematics' most romantic tales.

Évariste Galois

Galois was born near Paris in October 1811. His interest in mathematics began when, at the age of 14, he was forced to repeat his third-year classes at the *lycée* after failing his examinations. Mathematics, he found, helped stave off the boredom he felt for the rest of his schooling. Unfortunately, his growing passion for mathematics caused his schoolwork to deteriorate even further, and when he took the examination for entry into the prestigious École Polytechnique at the age of 15, he failed it and was forced to go to the less prestigious École Normale. It was there in the following year that he produced his first paper on mathematics: a competent although unremarkable piece of work on continued fractions. It was a promising start, but soon after, a series of unfortunate mishaps began that eventually led to his complete abandonment of the subject he loved with such passion.

Galois's next two papers (on polynomial equations) were rejected by the French Academy of Sciences. Worse, both manuscripts were unaccountably lost. Then, in July 1829, he failed once again to gain entry to the École Polytechnique, an occurrence that may have had a great deal to do with his answer to one particular question put to him by the examiner. When asked for an outline of the theory of "arithmetic logarithms," Galois answered (with total accuracy but demonstrating an amazing lack of tact and understanding) that there were no *arithmetic* logarithms. Following this disappointment, early in 1830 Galois presented yet another paper to the academy, this time in competition for the Grand Prize in Mathematics. The secretary, Joseph Fourier, took the manuscript home to read but died before he had made his report, and the paper was never found. This third loss of work Galois had submitted, coupled with his repeated failure to gain entry into the Polytechnique, drove him to reject the academic community, and he became what would nowadays be called a student radical. Within a year, he was expelled from school and was forced to try to support himself by giving private tutoring. Although he was not very successful in this, his mathematics continued to flourish, and it was during this period that Galois wrote what was to become his most famous paper, "On the Conditions of Solubility of Equations by Radicals," submitted to the academy in January 1831.

The submission of this paper was Galois's final attempt to have his work recognized. By March, when he had heard nothing from the academy, he wrote to the president to find out what had happened to his paper. Receiving no reply to that letter, he finally gave up. He would do no more mathematics. Instead he joined the National Guard (a Republican organization). But here he turned out to have no more luck than in mathematics. Soon after he had joined, the guard was disbanded following conspiracy charges. At a banquet held in protest on May 9, Galois proposed a toast to the king with an open knife in his hand—a gesture that his companions not unnaturally interpreted as a threat to the king's life—and the following day he was arrested. At his trial, he claimed that what he had actually said was "to Louis Philippe, *if he turns traitor*" but that the uproar had drowned out the last phrase. Whether or not this was true, he was acquitted and freed on June 15.

On July 4, Galois finally learned the fate of his paper to the academy. Rejecting it as "incomprehensible," the referee, Siméon Poisson, ended his report with these words:

> We have made every effort to understand Galois's proof. His reasoning is not sufficiently clear, sufficiently developed, for us to judge its correctness, and we can give no idea of it in this report. The author announces that the proposition that is the special object of this memoir is part of a general theory susceptible of many applications. Perhaps it will transpire that the different parts of the theory are mutually clarifying, are easier to grasp together rather than in isolation. We would then suggest that the author should publish the whole of his work in order to form a definitive opinion. But in the state that the part he has submitted to the Academy now is, we cannot propose to give it approval.

Connoisseurs of condescending rejections may feel that this report is a classic, and we have no idea whether or not its receipt had any influence on what Galois did next. But on July 14, he was arrested for appearing in public in the uniform of the now-disbanded National Guard, and he was sentenced to six months' imprisonment.

Shortly after his release on parole, Galois fell in love. But the *affaire* led to his early death, as somehow or other it resulted in his being challenged to a duel. (Alexandre Dumas suggested that the duel was really a disguised assassination plot, with political motives, but most historians of mathematics think this to be unlikely.) On May 29, the eve of the duel, Galois wrote a long letter to his friend Auguste Chevalier in which he outlined the current state of his mathematical theories, thereby providing the mathematical world with one final indication of the loss it was about to suffer. In the following day's duel (pistols at twenty-five paces) Galois was hit in the stomach and died twenty-four hours later.

What happened to his rejected paper? On July 4, 1843, Joseph Liouville addressed the French Academy, opening with these words: "I hope to interest the academy in announcing that among the papers of Évariste Galois I have found a solution, as precise as it is profound, of this beautiful problem: whether or not it is soluble by radicals. . . ."

The central concept that Galois had left to the world proved to be one of the most significant of all time, having applications in many fields of mathematics as well as in physics, chemistry, and branches of engineering. The concept is that of a *group*.

It is an entirely abstract concept. What makes it so powerful is the great number of examples of groups, examples of often quite different natures. Because of the versatility of the group concept, there are various ways of introducing it. The one I have chosen here uses symmetry properties of geometric figures in a plane, because it provides easily visualized examples. Later in this chapter we shall meet other types of groups.

Symmetry

Consider the isosceles triangle illustrated in figure 24. In everyday parlance, this geometrical figure is "symmetrical" about the vertical line in-

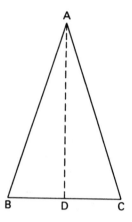

FIGURE 24   The symmetry of an isosceles triangle.

dicated. What we mean by the statement that the triangle ABC is *symmetrical* is that the part of the triangle to the left of the vertical line (the smaller triangle ABD) is the mirror image of that part to the right (ACD) with respect to an imaginary mirror placed along the vertical line AD, perpendicular to the plane. If we were to swap over (or *reflect*) the two halves of the figure, the result would be an entirely similar triangle in exactly the same position, but with the lines AB and AC interchanged and the line BC reversed.

In general terms, for any geometric figure S in the plane and any line *l* in the plane, the *reflection* of S in the *axis l* is the action of moving every point of S to its mirror image in *l*—to the point an equal distance from *l* along the line drawn through the point and perpendicular to *l*. Notice that it is the *action* of transforming the figure that is referred to as the *reflection*, not the result of that action. (For reasons that will become clear, we are concentrating on actions rather than their results.) The figure produced by applying a reflection to a figure S is called the *image* of S under that reflection. Figure 25 shows some examples of (the results of) reflections.

Using the notion of a reflection, the mathematician says that a figure S in the plane is *symmetrical* about an *axis of symmetry l* if the reflection of S in *l* results in an image that occupies exactly the same position in the plane as does S. (Figure 25[d] shows an example of symmetry about

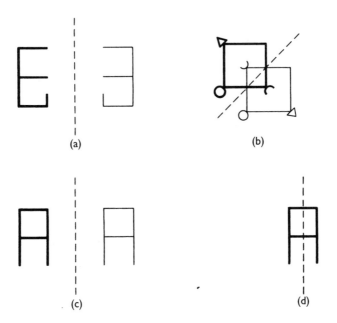

(a)   (b)

(c)   (d)

FIGURE 25 Reflections in an axis. In each case, the figure shown in **boldface** is reflected in the axis indicated by the dashed line, producing the lighter image. In (d), the image coincides with the original figure.

an axis.) Symmetry about an axis is sometimes referred to as *bilateral symmetry.*

Bilateral symmetry is what is meant by the everyday use of the word *symmetry* (for planar figures), but for mathematicians, there is another kind of symmetry, which is illustrated in figure 26. If the shape shown is rotated through an angle of 120° (in either direction) about the central point, it will end up occupying exactly the same position in the plane. This is an example of *rotational symmetry.*

Groups

Consider again the isosceles triangle in figure 24. How many symmetries does it have? That is, what are the reflections (in axes) and rotations

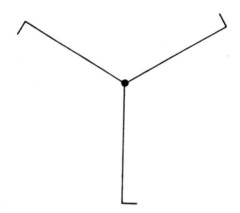

FIGURE 26   Rotational symmetry.

(about points) that send the triangle to an image occupying exactly the same position as the original triangle? There is reflection about the vertical line AD, and we will denote this reflection by the letter *r*. Are there any other symmetries? Obviously there are no other reflectional symmetries, but how about rotations? Certainly a rotation through 360° about any point will bring the figure back to its starting point, but this does not really count, for in this case the result would be no change at all (whereas in the case of the reflection *r*, there is a genuine change in that the points B and C end up occupying different positions from where they started). So we discount such trivial examples—or at least we almost do. But just as it is useful to consider the number 0 (which results in no change when we are adding) and the number 1 (which has no effect when we multiply), so too it proves useful to include as a symmetry the *identity transformation, I,* which leaves every point of the plane unchanged. Thus, *I* may be regarded as a rotation through 0°.

Suppose now that on the triangle ABC (see figure 27(a)), we perform the reflection *r* to produce the triangle ACB shown in figure 27(b). What happens when we perform *r* again, this time on ACB? We end up with the original configuration ABC again—figure 27(c). So the result of performing *r* twice in succession is the same as simply doing nothing (which we can express as the performance of the identity transformation *I*). We can express this symbolically by writing

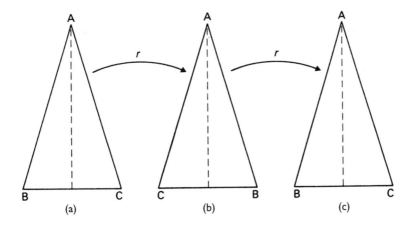

FIGURE 27 Successive reflections.

$$r * r = I,$$

where the asterisk * means "apply again." (So if *a* and *b* were two symmetries, *a* * *b* would mean the operation consisting of first applying *a*
and then applying *b* to the result.) Using the same notation, we may describe the (in this case trivial) effects of performing other sequences of
symmetries, thus:

$$r * I = r,$$
$$I * r = r,$$
$$I * I = I.$$

These four identities may be summarized in a table:

Isosceles triangle:

*	I	r
I	I	r
r	r	I

To see the effect of the application of a symmetry x followed by another symmetry y, look along the x row of the table until you come to the y column, and the entry you find is $x * y$, the result of the two symmetries combined.

Notice that our discussion implicitly assumed that the result of performing two symmetries in succession was itself a symmetry. This is indeed the case (you will see it is obvious if you think about it).

What happens when we perform the same analysis on the three-pronged shape shown in figure 26? Here there are three symmetries: a rotation clockwise through 120° (call it v), a rotation clockwise through 240° (call it w), and the identity I, where everything is left fixed. ("What about counterclockwise rotations?" you might ask. Well, a counterclockwise rotation through 120° gives the same result as w, and a counterclockwise rotation through 240° is equivalent to v, so we really have included all possibilities.) Since two successive rotations through 120° have the same effect as one rotation through 240° does, then

$$v * v = w.$$

Similarly, two 240° rotations are equivalent to one 120° rotation, so

$$w * w = v.$$

The full table of successive symmetries is

Tripod:

*	I	v	w
I	I	v	w
v	v	w	I
w	w	I	v

One more example. The equilateral triangle (see figure 28) has six symmetries. There is the identity, I, clockwise rotations v and w through 120° and 240°, respectively, and reflections x, y, z in the lines X, Y, Z, respectively. (The lines X, Y, Z stay fixed when the triangle

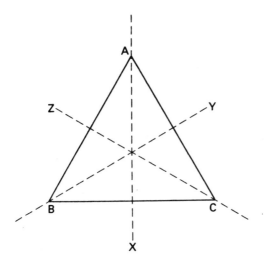

FIGURE 28 Symmetries of an equilateral triangle.

moves.) These symmetries combine as indicated by the following table:

Equilateral triangle:

*	I	v	w	x	y	z
I	I	v	w	x	y	z
v	v	w	I	z	x	y
w	w	I	v	y	z	x
x	x	y	z	I	v	w
y	y	z	x	w	I	v
z	z	x	y	v	w	I

If you wish to check these entries, you could try cutting out an equilateral triangle from cardboard, marking the corners A, B, C, and placing it on a sheet of paper on which the lines X, Y, Z are drawn. Then you can physically perform the various rotations and reflections. (You will have to mark your triangle on both sides to allow for reflections.)

What about combinations of more than two symmetries? There is no need to consider them, since the application of any number of symmetries is equivalent to a succession of pair combinations. For example, for the equilateral triangle just considered, $(x * y) * v$ is the same as $v * v$, which is just w, using the table twice. The parentheses were required to indicate the order in which the symmetries were to be combined: apply x followed by y, and then apply v to the result. The alternative grouping $x * (y * v)$ would mean apply x and then apply $y * v$ to the result. If you did things the second way, what would you get? Because $y * v$ is z, so $x * (y * v)$ is $x * z$, which is w, it would be the same result as was obtained the other way. If you think about it for a moment, you will realize that this is no accident. It is true for all combinations of symmetries: if a, b, c are symmetries of some figure, then

$$(a * b) * c = a * (b * c).$$

(This property is known as *associativity* of the operation *.)

We need to make one final observation before defining a "group." It is clear that if you apply any symmetry "backward," the result will be another symmetry. (For a reflection, there is no difference between the "forward" and the "backward" applications, of course. For rotations, you simply reverse the direction of the rotation.) The "backward application" of a symmetry x is called the *inverse* of x, often denoted by the symbol x^{-1} (which you should read as either "x to the minus one" or "x-inverse"). For the isosceles triangle, you get $r^{-1} = r$ (as for any reflection). For the tripod, you get $v^{-1} = w$ and $w^{-1} = v$, and for the equilateral triangle $v^{-1} = w$, $w^{-1} = v$, $x^{-1} = x$, $y^{-1} = y$, and $z^{-1} = z$. In all cases, $I^{-1} = I$. Do you notice anything about these results? If you check the tables in each case, you will notice that it is always true that $a^{-1} * a = a * a^{-1} = I$. Once again, this is no accident, and if you think about what is meant by the identity symmetry and the inverse of a symmetry, you will see why it happens.

By now you should have a vague feeling that something is very familiar about all of this—even if you have never thought about symmetries before. Doesn't it all look remarkably like the ordinary multiplication of rational numbers (fractions), especially if you exclude zero (which has no inverse)? The product of any two nonzero rational num-

bers is another rational number; grouping does not matter for multiplication—$(ab)c = a(bc)$ for any numbers a, b, c—and every nonzero rational number x has an inverse rational number x^{-1} ($= 1/x$) such that
$xx^{-1} = x^{-1}x = 1$, where that number 1 has the property that multiplication by it causes no change. Or perhaps this was not quite what you
were thinking of. Perhaps the example you had in mind was the whole
numbers (integers) with addition: the sum of two integers is another integer; $(a + b) + c = a + (b + c)$ always holds; there is an identity number 0 that causes no change when you are adding; and every integer x
has an inverse $(-x)$ such that $x + (-x) = (-x) + x = 0$. Or maybe you
had yet another example in mind. There are, in fact, many possibilities,
including the one Galois was thinking of when he analyzed the problem
of solving quintic equations. All these examples are special cases of Galois's general concept of a "group."

To the mathematician, a *group* consists of

1. a set G and

2. an operation $*$ that to any pair of elements x and y of G assigns
   an element $x * y$, which also belongs to G.

The operation $*$ is required to satisfy the following three conditions
("group axioms"):

3. It is associative: for any x, y, z in G,

$$(x * y) * z = x * (y * z).$$

4. There is an identity element I in G such that for any x in G

$$I * x = x * I = x.$$

5. Every member of G has an inverse: if x is in G, then there is an
   element y of G such that

$$x * y = y * x = I.$$

We have already met several examples of groups. If G is the set of all
symmetries of a given figure in a plane and $*$ is the operation of applying two symmetries in succession, the result is a group. Or if G is the set

of nonzero rational numbers and $*$ is ordinary multiplication, the result is a group. Or if G is the set of integers and $*$ is ordinary addition, the result is a group. Both these examples of number groups have operations that are *commutative*; that is,

$$x * y = y * x$$

for all elements x, y of the group. But this need not necessarily hold for groups in general. For instance, it is not true for the equilateral triangle group. If you look back at the table for this group, you will see that $x * v = y$, but $v * x = z$. Groups in which $*$ is commutative are often referred to as *Abelian* (after the Norwegian mathematician Abel, mentioned earlier).

There is no need for the operation $*$ of a group to be a familiar operation, say an arithmetical one. But if it is, then it is customary to call it by its usual name (multiplication, addition, or whatever it may be). If the operation is unknown or has no common name, it is customary to refer to $*$ as the *multiplication* of the group, and to $a * b$ as the *product* of a and b in the group. This is purely a convenience, and on no account should any inference be drawn from such usage.

Because there are so many different examples of groups, the group concept is a very powerful one—not only in mathematics (where groups crop up all over the place) but in other subjects as well. The periodicities of a crystal, the symmetries of atoms, and the interactions of elementary particles all involve groups. Anything that can be said about groups in general is true for any particular group. But just how do mathematicians establish the properties of an *arbitrary, abstract* group? The answer is that everything must be done by means of rigorous mathematical proofs, starting from the definition of a group.

As an example, we shall prove that any element of a group must have *exactly one* inverse. (This must be established before the notation x^{-1} can be used for the inverse of x.) Condition 5 in the definition of a group guarantees that every element of a group has *at least one* inverse, but it does not preclude the existence of more. Of course, for each of the examples of groups given earlier, it is obvious that no element has more than one inverse, but this does not help us here. What is required is a proof that will apply to all cases, including possible examples of groups that we may not have considered before.

Here, then, is the proof. Let G be any group, and let x be any member of G. Let y and z be two inverses of x. The aim is to prove that $y = z$. Being inverses of x, both y and z satisfy the requirement of condition 5:

$$x * y = y * x = I, \qquad (3)$$

$$x * z = z * x = I. \qquad (4)$$

Applying condition 4 to y gives

$$y = I * y.$$

So, using equation (4),

$$y = (z * x) * y.$$

Using condition 3, it follows that

$$y = z * (x * y).$$

So, using equation (3),

$$y = z * I.$$

Thus, applying condition 4 to z gives

$$y = z.$$

This completes the proof. Notice that all the structural conditions imposed on a group by the definition are required to carry out this proof.

As you might imagine, most proofs in *group theory* are more complicated than this rather easy example (and consequently are not usually written out in quite so much detail). Often they involve other concepts, but they always share the characteristic of consisting entirely of logical steps based only on the initial assumptions.

More Examples of Groups

All the symmetry groups we just considered were connected with plane figures, but the same ideas apply to solid shapes of three dimensions. A cube, for example, has 24 rotational symmetries (rotation about an axis now, not about a point, as in two dimensions). To see this, notice that any vertex of a cube may be moved to any other and that the edges leading to that vertex may be rotated in three ways. If reflections are included (reflections in a *plane*, not a line) the cube has a total of 48 symmetries.

A *dodecahedron,* which consists of 12 identical regular pentagons fitted together to form a ball-like solid (see figure 29), has 60 rotational symmetries (120 symmetries when reflections are included). In both the cube and the dodecahedron, the rotational symmetries on their own form a group that "sits inside" the group of *all* the figure's symmetries. Mathematicians would say that the rotational symmetries form a *subgroup* of the entire symmetry group in each case.

The rotational symmetries of a dodecahedron form the smallest non-commutative *simple* group, and it was the simplicity of this group, together with the fact that it has a nonprime number of elements, that Galois used to show that the general quintic polynomial equation could not be solved by radicals.

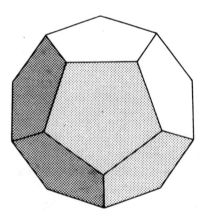

FIGURE 29   A dodecahedron.

So far we have seen examples of groups that are infinite (e.g., addition of the integers) and others that are finite (e.g., the various symmetry groups). Most of our attention in this chapter will now be focused on finite groups. Matrices provide examples of both finite and infinite groups.

A *matrix* is just a rectangular array of numbers (which may be either rational or real as far as our examples are concerned), usually drawn enclosed in brackets, for example,

$$\begin{bmatrix} 1 & 3 & \frac{3}{4} \\ 2 & \frac{1}{4} & -5 \end{bmatrix}.$$

They can be any size, but we shall be concerned exclusively with *square matrices,* in which the number of rows is the same as the number of columns (that number being known as the *order* of the matrix). Thus an example of a (square) matrix of order 2 is

$$\begin{bmatrix} 21 & -5 \\ 3{\cdot}8 & 20 \end{bmatrix}.$$

Matrices have their own arithmetic. The rule for adding two matrices (of the same order) is straightforward: you simply add corresponding entries. Thus

$$\begin{bmatrix} 1 & 3 \\ -2 & 6 \end{bmatrix} + \begin{bmatrix} 2 & 5 \\ 3 & 1 \end{bmatrix} = \begin{bmatrix} 3 & 8 \\ 1 & 7 \end{bmatrix}.$$

Multiplication is a little more complicated. Briefly, you multiply the rows of the first matrix by the columns of the second, term by term, adding the answers as you go. For matrices of order 2, this is perhaps best explained by means of an algebraic example followed by a numerical one:

$$\begin{bmatrix} a & b \\ c & d \end{bmatrix} \times \begin{bmatrix} v & w \\ x & y \end{bmatrix} = \begin{bmatrix} (av + bx) & (aw + by) \\ (cv + dx) & (cw + dy) \end{bmatrix},$$

$$\begin{bmatrix} 1 & 3 \\ -2 & 5 \end{bmatrix} \times \begin{bmatrix} 2 & 4 \\ 3 & 1 \end{bmatrix} = \begin{bmatrix} (2 + 9) & (4 + 3) \\ (-4 + 15) & (-8 + 5) \end{bmatrix}$$

$$= \begin{bmatrix} 11 & 7 \\ 11 & -3 \end{bmatrix}.$$

A second example shows that matrix multiplication is not commutative (although addition is):

$$\begin{bmatrix} 2 & 4 \\ 3 & 1 \end{bmatrix} \times \begin{bmatrix} 1 & 3 \\ -2 & 5 \end{bmatrix} = \begin{bmatrix} (2-8) & (6+20) \\ (3-2) & (9+5) \end{bmatrix}$$

$$= \begin{bmatrix} -6 & 26 \\ 1 & 14 \end{bmatrix}.$$

Why such a complicated definition of multiplication? you might ask. Why not simply multiply the corresponding entries, much as in addition? The point is that mathematicians developed and studied matrices with certain applications in mind (in particular, the solution of large systems of simultaneous linear equations) and those applications required the definitions given. Matrix arithmetic is so important nowadays that any computer system aimed at scientific or commercial users is supplied with software to handle it as a matter of course. Indeed, matrix arithmetic is probably the numerical task most often performed by present-day computers.

The definitions of addition and multiplication of matrices of order 3 or higher are similar to those for order 2 given earlier, and practically everything said in what follows applies (with only the obvious modifications) to matrices of any size, but for clarity we shall continue to concentrate on matrices of order 2.

The matrices of order 2 (or indeed of any order) form a group in regard to addition. The sum of two matrices (of order 2) is a matrix (of order 2); the addition is associative; there is an identity matrix

$$\begin{bmatrix} 0 & 0 \\ 0 & 0 \end{bmatrix},$$

and the inverse of any matrix is obtained by putting a minus sign in front of all its entries. This group is in fact commutative.

How about multiplication—does that give rise to a group? Certainly the product of two matrices (of order 2) is again a matrix (of order 2), and matrix multiplication is associative. (This may not be clear immediately, but if you work it out algebraically, you will see that it is true.) There is an identity element, namely, the matrix

$$\begin{bmatrix} 1 & 0 \\ 0 & 1 \end{bmatrix},$$

that leaves any matrix unaltered when multiplied by it. Turning to the question of inverses, by straightforward calculation you can check that

$$\begin{bmatrix} a & b \\ c & d \end{bmatrix} \times \begin{bmatrix} d/H & -b/H \\ -c/H & a/H \end{bmatrix} = \begin{bmatrix} 1 & 0 \\ 0 & 1 \end{bmatrix},$$

where $H = ad - bc$, and similarly with the two matrices on the left-hand side swapped around. So the matrix

$$\begin{bmatrix} a & b \\ c & d \end{bmatrix}$$

has as its inverse the matrix

$$\begin{bmatrix} d/H & -b/H \\ -c/H & a/H \end{bmatrix},$$

provided this second matrix exists! The only thing that can go wrong is if the quantity H is zero. (Remember that division by zero is never possible.) Matrices for which this number H is nonzero do have inverses. Such matrices are called *invertible* (or sometimes *nonsingular*). Matrices for which the number H is zero do not have inverses and are called *singular* (or *noninvertible*).

Because of the existence of singular matrices, which have no inverses, the matrices do not form a group under multiplication. But if you consider only the invertible matrices, you do get a group. Is this evident? Not quite. Associativity is not a problem, since it holds for all matrices, whether invertible or not. Since the identity matrix is invertible (and hence a member of the collection being considered), there is an identity element. Inverses? Of course, all the members of the chosen collection do have inverses, but you must check that these inverses *themselves lie in the collection.* This is an easy matter, but it must be done. There is one final thing to check. You must know that when two invertible matrices are multiplied together, the resulting matrix is itself invertible (i.e., still in the collection). Again, this is simply a matter of checking the algebra. (Try it for yourself.)

The invertible matrices form a group that is infinite. (It is not commutative, as indicated earlier.) Some very important groups we consider later in this chapter consist of finite subgroups of the invertible matrices.

Another important class of groups is the *clock groups,* the most common example being the familiar twelve-hour clock. Take as the set the integers from 1 to 12, with the "clock addition" operation, where 12 counts as a "zero" and counting past 12 takes you back to the beginning again. So, for example, 5 plus 5 is 10, 7 plus 8 is 3, 11 plus 11 is 10, and 7 plus 12 is 7. This produces a group in which the identity element is 12 and the inverse of any member of the group is the difference between it and 12; thus 7 is the inverse of 5, 9 the inverse of 3, and so on.

There is nothing special about the number 12 here. Any number will do. The clock group of order 10 is the structure that lies behind the decimal number system; the clock group of order 24 corresponds to the twenty-four-hour clock; the clock group of order 60 is connected with the measurement of time; and the clock group of order 360 is related to the measurement of angles.

The mathematician's name for a clock group is a *cyclic group,* so named because the elements of such a group go round in cycles (like a clock). For instance, the cyclic group of order 3 has the following table:

+	1	2	3
1	2	3	1
2	3	1	2
3	1	2	3

Incidentally, you may have noticed that the clock groups are just a special case of the mod arithmetic mentioned in chapter 1.

Simple Groups

One of the primary aims in any branch of science is to identify and study the "basic objects" from which all other objects are constructed. In biology, these are the cells (or possibly the molecules); in chemistry, the atoms; and in physics, the fundamental particles (currently the quarks). So too in many branches of mathematics. The classic example is number theory, in which (according to the fundamental theorem of arithmetic described in chapter 1) the prime numbers are the basic building blocks. In each of these examples, the basic objects of the theory are *structurally simple,* in the sense that (from the point of view of the theory) they cannot be decomposed into smaller entities of the same kind. (The atoms cannot be broken apart by chemical means; the prime numbers cannot be broken apart by division; and so on.)

The fundamental building blocks in group theory are the *simple groups.* To explain what these are and how any given (finite) group can be split into component simple groups, we require the concept of a *telescopic image* of a group. Roughly speaking, when you make a telescopic image of a group G, you get a sort of "scaled-down" version of G. The group operation $*$ of G is reflected in the telescopic image, although in a reduced form. It is a bit like looking at an object through the wrong end of a telescope: the main features of the object are preserved, but it appears smaller, and many features may no longer be distinguishable.

To be a little more precise, if you start with a group G, then to form a *telescopic image*, G', of G you must associate with each element a of G an element a' of G' (called the *image* of a) in such a way that for any pair of elements a, b of G with images a', b', respectively, the product of a' and b' in G' must be the image of the element $a * b$ of G. (Thus G' preserves the structure of G.) There is nothing here to prevent several elements of G having the same image in G', and indeed, it is this "collapsing" or "coalescing" of elements that accounts for the reduction in size when you go from G to G'. Mathematicians refer to telescopic images as *homomorphic images.*

Every group G has at least two telescopic images. One of these is G itself, for which each element of G is its own image. This situation trivially satisfies the requirements for being a telescopic image, although it clearly is an extreme case. The other telescopic image that any group G possesses lies at the other extreme. It is the *point image* of G, namely, the group that has only one element, the identity element e. Notice that the definition of a group allows for an identity element to form a group on its own, albeit a rather trivial one, with the multiplication "table"

$$e * e = e.$$

In the point image, every element of G has the same image, namely, e, so in this case also, the requirements of a telescopic image are satisfied.

Clock groups provide examples of groups that have other telescopic images besides the two trivial ones just mentioned. For instance, let G be the clock group of order 24, and let G' be the clock group of order 12. For each number n from 1 to 12 in G, its image n' is n itself. For n between 13 and 24 (inclusive), n' is $n - 12$. Then G' is a telescopic image of G. For example, take the elements 7 and 18 in G. Their images in G' are 7 and 6, respectively. The sum of 7 and 6 in G' is 1. According to the definition of a telescopic image, this should be equal to the image of the sum of 7 and 18 in G. Well, in G, 7 and 18 have the sum 1, and the image of 1 in G' is indeed 1. (Notice that the process of going from G to G' is just the usual method of converting time from the twenty-four-hour clock to the twelve-hour clock.)

Observe that in this example it was important that 12 (the order of G') divides 24 (the order of G). A clock group of prime order has no telescopic images other than itself and the point image. This provides a family of examples of the central notion of this chapter:

> A *simple group* is a group whose only telescopic images are itself and the point image.

By means of telescoping, every finite group may be split into a unique set of simple groups in much the same way that a composite number is decomposed into its prime factors (see chapter 1). Indeed, the analogy goes further: the number of elements in each of these simple-group components is a factor of the number of elements of the original group, and the product of all these numbers is equal to the number of elements in the original group. There, however, the analogy stops. For one thing, the simple-group components of a group may contain a composite number of elements. (As mentioned earlier, the rotational symmetries of a regular dodecahedron form a simple group of order 60.) Also, whereas the product of all the prime numbers in a given set is a unique number, a given set of simple groups can often be combined in different ways to form distinct groups.

The Classification Problem

Having identified the simple groups as being the "fundamental particles" of finite group theory, mathematicians tried to impose some form of classification on the simple groups. Roughly speaking, they wanted to be able to determine which groups are simple and which are not. Of course, the definition of simplicity itself gives some form of answer: the simple groups are those that have only two telescopic images. But that is not the sort of answer they wanted. An acceptable solution would describe the actual structures that, considered as groups, turned out to be simple. This could be in the form of a general pattern, giving rise to a family of groups, or a description of an individual "one-off" group.

For example, one easy result is that all clock groups of prime order are simple (and clock groups of composite order are not). In fact, these are the only examples of commutative simple groups, so already we have a complete classification of all the *commutative* simple groups (in the form of a "regular" family). It was the classification of the noncommutative simple groups that took all the effort. Beginning around the 1940s, when mathematicians began to work toward the classification theorem (though not necessarily with that goal in mind), they discovered several infinite "regular" families of simple groups. In the end, they found a total of eighteen such families, including the family of all clock groups of prime order just mentioned and another easily described family we'll encounter presently. Also found were a number of highly irregular "oneoff" groups that did not fit into any known pattern. The first five of these strange *sporadic* simple groups, as they came to be called, had been found by Émile Mathieu in the 1860s. The smallest of Mathieu's groups has exactly 7920 elements, and the largest, 244,823,040. It was not until a century later, in 1965, that a sixth sporadic group was discovered by Zvonimir Janko. Janko's group, which has 175,560 elements, consists of a certain collection of matrices of order 7 (with matrix multiplication as the group operation). The way in which this group was discovered is indicative of the kind of work that led to the discovery of an eventual collection of twenty-six sporadic simple groups. It arose as a result of examining the seventeenth of the regular families of simple groups, a family that was discovered by Rimhak Ree in 1960 and nowadays is known as the Ree family.

With every simple group are associated certain smaller groups that provide information about the structure of the simple group. Called *centralizers of involution,* their precise definition will be given later, but for the Ree groups, they consist of matrices of order 2 whose entries come from a finite set of numbers of size equal to some odd-numbered power of 3. (If the odd power of 3 is 1, the finite set of numbers turns out to be just 1, 2, and 3.) As part of an attempt to prove an early, restricted form of classification theorem, it was necessary to show that the Ree groups are the only simple groups whose associated centralizers of involution consist of matrices of order 2 whose entries come from a finite set of numbers of size equal to an odd power of some prime number p. The evidence suggested that the prime number p here had to be 3, and this was eventually proved to be true, except in the single case in which p is 5 and the odd power is 1. It was this exceptional case that Janko set out to investigate. His intention was to eliminate this one remaining obstacle by proving that there are no simple groups of the type concerned in which the number set involved has size 5^1 (i.e., 5). He did not succeed in this goal, but he did manage to obtain a rather curious result. He proved that if there were such a simple group, it must contain exactly 175,560 elements. Such a precise result suggested that there had to be an actual group hidden somewhere in the background, and after an immense amount of hand calculation, Janko succeeded in finding it. Thus was the sixth sporadic group discovered. In honor of Janko, it was named J_1.

By using similar techniques with families of simple groups other than the Ree family, Janko soon found evidence for the existence of two more sporadic simple groups, one with 604,800 elements and the other with 50,232,960. He could not, however, actually find such groups. The smaller of the two, J_2, was eventually found by Marshall Hall Jr. and David Wales, and the larger one, J_3, was rooted out by Graham Higman and John McKay (using a computer to perform the calculations).

In a more or less similar fashion, in the following years, several more sporadic groups were discovered until, in 1980, the last of the twenty-six such "one-off" groups was constructed (its existence having been suspected since 1973) by Robert Griess. It is by far the largest of the sporadic groups, a fact that earned it its name of "the Monster." For the record, the number of elements in the Monster is

808,017,424,794,512,875,886,459,904,961,710,757,005,
754,368,000,000,000.

(That is roughly 8 followed by 53 zeros.) It consists of a certain collection of matrices (with entries taken from the complex numbers) of order

196,883. We should add that Griess performed all the necessary calculations to determine the Monster by hand. The fact that the group was "cooperative" enough to allow such an approach to succeed caused Griess to rename his group "the Friendly Giant."

The discovery of the Monster was one of the last steps in the proof of the classification theorem. It is now known that the finite simple groups consist of the groups that make up the eighteen regular, infinite families of groups (the first of which is the family of clock groups of prime order), together with the twenty-six sporadic groups, *and no more*. This is the result that took up 500 articles and almost 15,000 pages in mathematical journals.

The Eighteen Families and the Odd Ones Out

Very often in mathematics, progress is made by formulating a theorem and then proving it. With the classification theorem, however, this was not the case. Until this theorem was proved, there was no way of knowing even the size of the problem. There might, for instance, have been many more than twenty-six sporadic groups, possibly even infinitely many, which would have meant that the goal being sought could never be achieved. The greater part of the work was done on the basis of "let's find out about simple groups" rather than as a determined push toward a stated theorem. This makes it difficult to say exactly when the work that led to the final classification began. In his address to the 1954 International Congress of Mathematicians in Amsterdam, Richard Brauer proposed a method for trying to classify the simple groups (of even order, although this restriction ultimately proved to be redundant), and this could count as one starting point. Another, possibly less debatable, starting date would be 1972, the year in which Daniel Gorenstein gave a series of lectures at the University of Chicago outlining a sixteen-step program that should lead to the eventual solution of the classification problem. In some respects, the final assault was made possible by a key result obtained by Walter Feit and John Thompson in 1962, and this too could be regarded as "the beginning of the end." At any rate, before saying any more, we first should say something about the nature of the groups that appear in the complete solution.

The first of the eighteen regular families has already been mentioned: the family of all clock groups of prime order. The second family may likewise be easily described. For any whole number n greater than 4, the group of all even permutations of n symbols is a simple group, and the collection of all such groups forms the second family. What is an *even permutation of n symbols?* Consider $n = 4$ (the first interesting case). Take four symbols, say the letters A, B, C, D. In alphabetical order, these letters form a "word," ABCD. By repeatedly swapping pairs of letters, it is possible to rearrange these four letters into any one of $4 \times 3 \times 2 \times 1 = 24$ different "words" (or orders). Any such rearrangement is called a *permutation* of ABCD. It is an *even* permutation if it is obtained by an even number of swaps, and an *odd* permutation if the number of swaps is odd. For instance, CBDA is an even permutation, being obtained from ABCD by first swapping A and C and then swapping A and D, whereas BACD is an odd permutation, being arrived at from ABCD by swapping just the single pair A and B.

Happy so far? Now we shall do what we did with symmetries, and think of permutations not as the final ordering of the letters but, rather, as the sequence of swaps that achieves that final ordering. This means that we can think of *combining* two permutations to form a single permutation: if a and b are permutations (i.e., sequences of swaps) of ABCD, then $a * b$ is the permutation that consists of first performing the a swaps and then the b swaps. For example, if a swaps A and C and then C and D, and if b swaps A and B, then, starting from ABCD, you get

a transforms ABCD to DBAC,

b transforms DBAC to DABC,

$a * b$ transforms ABCD to DABC.

Obviously, the operation $*$ is associative. The identity permutation e, which changes nothing, acts as an identity operation; that is,

$$a * e = e * a = a, \quad \text{for any } a.$$

And the inverse of any permutation obviously consists of the same swaps performed in the opposite order, so if a swaps A and C and then C and D, then a^{-1} swaps C and D and then A and C. (Check for yourself that $a * a^{-1} = a^{-1} * a = e$.) Thus the permutations of the four letters A, B, C, D constitute a group. The even permutations on their own also form a group, a subgroup of the group of all permutations consisting of exactly half the elements. (This is a simple consequence of the fact that the sum of two even numbers is again an even number, so the * product of two even permutations is another even permutation.) This smaller group is called the *alternating group* of degree 4. It is an exact copy of the group of all the rotational symmetries of a regular tetrahedron (i.e., both groups have the same table and so, to all intents and purposes, are the same group).

So much for $n = 4$. The same considerations lead to an alternating group of degree n for any n greater than 2. (If there are only two symbols, there is only one nontrivial permutation, which is odd, so the corresponding alternating group would be the one-point group.) For $n = 3$, the permutation group has $3 \times 2 \times 1 = 6$ elements, and the alternating group is a mirror image of the clock group of order 3. (If a is the even permutation of ABC that swaps A and B and then A and C, then a sends ABC to BCA and $a * a$ sends ABC to CAB, and these are the only even permutations there are besides the identity. (In particular, $(a * a) * a$ takes ABC to ABC, and the clock has traveled around once.)

For any n greater than 4, the alternating group of degree n is a simple group. It is the simplicity of the group that lies behind the insolubility (by radicals) of all polynomial equations of degree greater than 4. But wait a moment. Didn't I say earlier that it was the simplicity of the group of rotational symmetries of a regular dodecahedron that led Galois to his conclusion about quintics? Indeed I did. The dodecahedron group is an exact copy of the alternating group of degree 5.

After the prime clock groups and the alternating groups of degree greater than 4, the remaining sixteen regular families are harder to describe in an account like this. They all are groups of matrices of appropriate sizes. In some cases, the families were first described in terms of the matrices involved; in others, the family was first defined in other terms and only after considerable effort was a matrix description obtained.

What about the way in which the classification problem was solved? By the twentieth century, many of the regular families were known, as were Mathieu's five sporadic groups. From the observation that all the known noncommutative simple groups contained an even number of elements, William Burnside conjectured that the same would be true of all noncommutative simple groups, however many there were and whatever other properties they might have. In 1962, Burnside's conjecture was proved correct by Walter Feit and John Thompson of the University of Chicago, a result for which they were awarded the Cole Prize in Algebra in 1965. Giving a foretaste of the extreme length of the eventual proof of the complete classification theorem, the proof of the Feit-Thompson theorem filled an entire 255-page issue of the *Pacific Journal of Mathematics*. (Mathematical journals like that one typically contain twenty or thirty papers on widely differing topics.)

With the Feit-Thompson theorem available, the way was suddenly open for an advance toward the classification theorem along the lines outlined by Brauer in his 1954 lecture mentioned earlier. The problem, you see, has two aspects. One is to identify the simple groups (or families thereof) required for the classification. Aside from one remaining family and a few sporadic groups, this step had been accomplished by 1960 (although at the time, this was not clear). The other step is to prove that every simple group does indeed fall into one of the given categories, which is where much of the proof's complexity lies. The problem is that you must start with an arbitrary simple group (i.e., all you know is that it *is* a simple group) and somehow show that it is (an exact copy of) a member of one of the regular families or else one of the listed sporadic groups. What Brauer suggested was a way of attacking this problem.

His idea was to concentrate on the elements a of the group (other than the identity element e) for which $a * a = e$. Such group elements are called *involutions,* and it is easy to show that any group with an even number of elements must contain at least one involution. (Try this yourself. All you need to know about groups is the definition given earlier. The solution is so neat and concise that its discovery is well worth the effort it might take you to find it.) According to the Feit-Thompson theorem, it follows that every noncommutative simple group contains involutions.

What Brauer did (and remember, this was before the Feit-Thompson theorem but after Burnside had made the conjecture that became their theorem) was to calculate centralizers of involutions in several of the known regular families. Centralizers? The *centralizer* of an element g of a group G is the set of all elements a of the group for which $a * g = g * a$. If G is commutative, the centralizer of any element is just G itself, of course, but in other cases, this need not be so. What is true—and is quite straight-forward to verify—is that the centralizer of any element of G is a subgroup of G. Brauer's observations were encouraging. All the centralizers-of-involutions groups had the same general structure as the original simple group, though in an embryonic form. This led Brauer to suspect that it might be possible to reconstruct the entire group from a knowledge of these centralizer groups, and in certain special cases, he was able to confirm this suspicion.

Not only does Brauer's work lie behind the discovery of many of the sporadic groups (the three Janko groups have been mentioned already), but it provides the beginning of a way of pinning down an arbitrary given simple group within the proposed classification. First, show that the centralizer of an involution in the given group closely resembles the centralizer of an involution in one of the known simple groups in the classification scheme. Then try to extend this highly localized "tie-up" to a complete equivalence. This last step is by no means easy; the centralizer of an involution is just a tiny part of the entire group, so it is a bit like trying to deduce the theme of an entire jigsaw puzzle from just one piece.

Following Brauer's approach was a part of the sixteen-step program outlined by Gorenstein in his 1972 Chicago lectures. Gorenstein himself thought that the task could be accomplished by the end of this century. Most of his audience thought this was wildly optimistic, but they all reckoned without the presence in the audience of a young mathematician who had just completed his graduate studies. Starting with a key result known as the *component theorem,* Michael Aschbacher of the California Institute of Technology (Caltech) stormed his way through the problem, proving one startling result after another, with the result that by 1980, just eight years after Gorenstein's lectures, Solomon was able to supply the one small step for mathematics that marked the completion of the proof. (Aschbacher won the 1980 Cole Prize for Algebra for his work.)

Along the way, the various remaining sporadic groups were uncovered. As with all the regular families except the first two, these all consist of certain sets of matrices, and in some cases, computers were required to perform the necessary computations. Given that these twenty-six groups are, by their very scarcity among the infinitude of all simple groups, quite special, it should come as no surprise to learn that they have connections with other branches of mathematics. For instance, the discovery in 1968 by John Conway of the three sporadic groups that now bear his name was based on Leech's lattice, a mathematical structure that arose in work on the design of error-correcting codes (methods for encoding transmitted information so that distortions and occasional losses can be compensated for). Two of Mathieu's sporadic groups are related to the Golay errorcorrecting code (named after its inventor, the mathematician Marcel Golay) that is often used for military purposes. Connections such as these provide some reason for interest in the classification theorem, but its main "claim to fame" outside group theory itself undoubtedly lies in the incredible length of the proof. Looking back on the proof soon after its completion in 1980, Michael Aschbacher wrote:

> Much of the mathematics involved has been produced only recently and will no doubt be improved once there is time for the techniques to sink in. Still, it is hard to imagine a short proof of the theorem. I personally am skeptical that a short proof of any kind will ever appear.
>
> Long proofs disturb many mathematicians. For one thing, as the length of a proof increases, so does the possibility of an error. The probability of an error in the most recent proof of the classification theorem (see momentarily) is virtually 1. On the other hand the probability that any single error cannot be easily corrected is virtually zero, and as the proof is finite, the probability that the theorem is incorrect is close to zero. As time passes and we have an opportunity to assimilate the proof, that confidence level can only increase.
>
> It is also perhaps time to consider the possibility that there are some natural, fundamental theorems that can be stated concisely, but do not admit a short, simple proof. I suspect that the classification theorem is such a result. As our mathematics becomes more sophisticated, we may encounter such theorems more frequently.

Putting It All Together

While the Enormous Theorem may indeed admit of no "short" proof, the 1980 completion of the quest to classify the finite simple groups presented the challenge of finding a single, coherent proof, even if it required several volumes to contain it. In the early 1990s, a number of group theorists began the lengthy task of trying to produce just such a proof. Led by Richard Lyons and Ronald Solomon, the team set out to produce what they called a "second-generation" proof of the theorem, one that would fit nicely onto a single shelf of a bookcase.

To date, they have produced the first two volumes of what is likely to be a series of a dozen or more books. With the series unlikely to become a mathematics best-seller, the work is being done with posterity in mind, and both volumes have been published by the American Mathematical Society. Volume 1, which weighs in at a slim 165 pages, is little more than an outline of what is to follow. Volume 2 starts the proof in earnest.

During the course of producing a single proof, it is likely that simplifications will be discovered, along with side results. Indeed, that has already happened on several occasions. However, it is thought unlikely that there will be any major breakthroughs leading to, say, a single-volume proof. Instead, any mathematician of the mid-twenty-first century who wants to understand why the finite simple groups fall into four categories—cyclic, alternating, Lie-type, and the twenty-six sporadic groups—must be content with working through a single 5000-page account. The groups may be simple, but the proof is not. It's that simple.

Suggested Further Reading

A gentle introduction to group theory can be found in chapter 7 of *Concepts of Modern Mathematics,* by Ian Stewart (Pelican, 1981). A more complete treatment can be found in the book *A Course in Group Theory,* by John F. Humphreys (Oxford University Press, 1996).

At a higher level, a detailed account of the subject of this chapter is given in *Finite Simple Groups,* by Daniel Gorenstein (Plenum, 1982).

The definitive account of the classification of the finite simple groups is contained in *The Classification of Finite Simple Groups,* vols. 1 and 2, by Daniel Gorenstein (Plenum, 1982).

6

Hilbert's Tenth Problem

A Historic Gathering

In August 1900, the world's best mathematicians gathered in Paris for the Second International Congress of Mathematicians (an event that, except for periods of war, has continued to be held every four years at different venues around the globe). Among them was a 38-year-old professor from the University of Göttingen, David Hilbert. As one of the leading mathematicians of the time, Hilbert was scheduled to deliver one of the keynote addresses to the meeting. The day fixed for his lecture was August 8.

Because the meeting was being held on the eve of the twentieth century (indeed, it had been moved up a year so that it could), Hilbert chose to use his lecture not to look back over some recent work (which is the usual format for such talks) but, rather, to point the way toward the future.

"We hear within us the perpetual call: there is the problem," he announced. "Seek its solution. You can find it by reason, for in mathematics there is no *ignorabimus* [we shall not know]." To emphasize this call, he presented to the meeting not one but a list of twenty-three major unsolved problems—problems whose solutions, if found, would each mark a significant advance in mathematical knowledge. Most of these problems were (or became) known by specific names, such as the *con-*

tinuum problem (the first on Hilbert's list—see chapter 2) or the *Riemann problem* (see chapter 8), but one in particular became universally known by its position on Hilbert's list: the tenth.

Hilbert's tenth problem has its origins in an algebra textbook, the *Arithmetica,* written around A.D. 250 by Diophantus of Alexandria (see chapter 10). In accordance with the kinds of problems considered in that tract, present-day mathematicians use the name the *Diophantine equation* to refer to any equation in one or more variables, with integer coefficients, where a solution is sought *consisting entirely of integers.* It is this last condition that makes the mathematics of Diophantine equations radically different from the solution of equations in the real numbers (or possibly the complex numbers). (Actually, the nomenclature is a bit confusing at first. The adjectival use of the word *Diophantine* refers not to the equation so much as to the kind of solution sought. Thus the equation

$$3x^2 - 5y^2 + 2xy = 0$$

is referred to simply as an "equation" if real-number solutions are sought but as a "Diophantine equation" if only integer solutions are wanted.)

Solving Diophantine equations is quite different from solving the same equations in real numbers. For instance, if the equation

$$x^2 + y^2 = 2 \tag{5}$$

is regarded as a regular equation for real numbers, it will have infinitely many solutions. Given any real number r between $-\sqrt{2}$ and $+\sqrt{2}$, if

$$s = +\sqrt{2 - r^2},$$

then $x = r$, $y = s$ gives you a solution. Considered as a Diophantine equation, however, there are just four solutions:

$$x = +1, \quad y = +1; \quad x = +1, \quad y = -1;$$
$$x = -1, \quad y = +1; \quad x = -1, \quad y = -1.$$

If you change the equation just slightly, say to

$$x^2 + y^2 = 3, \qquad\qquad (6)$$

there still are infinitely many real solutions but no integer solutions. As a Diophantine equation, equation (6) cannot be solved. So what is the difference between equations (5) and (6)? More generally, is there a way of telling whether or not a given Diophantine equation can be solved? For instance, would it be possible to write a computer program that, given a Diophantine equation, indicates whether it has a solution? This is essentially the question Hilbert asked as the tenth in his list of problems. Its solution in 1970 by the 22-year-old Russian mathematician Yuri Matyasevich came only after a great deal of work had been done on the problem, stretching back to the 1930s and encompassing results in mathematical logic, computation theory, and algebra.

Diophantine Equations and the Euclidean Algorithm

The simplest kind of Diophantine equation is a linear one, with one unknown. In fact, the only information we have about Diophantus' life is itself in the form of such an equation. A fourth-century problem states that his boyhood lasted for one-sixth of his life, that his beard grew after another one-twelfth, that he married after one-seventh more, and that his son was born five years later and lived to half his father's age, dying four years before him. If you let x denote the age at which Diophantus died, this information gives us the equation

$$\frac{1}{6}x + \frac{1}{12}x + \frac{1}{7}x + 5 + \frac{1}{2}x + 4 = x,$$

which has the solution $x = 84$. (Strictly speaking, this is not a Diophantine equation, since the coefficients are not integers, but if you multiply through by the least common multiple of all the denominators in the coefficients, you will obtain an equation with integer coefficients.) Whether or not Diophantus really did live to age 84, the fact remains that the solution of a linear Diophantine equation in one unknown is a trivial matter. The equation

$$ax = b$$

has an integer solution if, and only if, a divides b (exactly), in which case the solution is the integer b/a. This condition is so simple that it is easy to write a computer program that will tell you at once whether such a Diophantine equation has a solution.

What about linear Diophantine equations in two unknowns? Here again, there is a simple way of finding out whether there is a solution. To see if the equation

$$ax + by = c$$

has an integer solution, first calculate the highest common factor, d say, of a and b. If d divides c, then there is a solution; if d does not divide c, there is no solution.

For example, does the equation

$$6x + 15y = 12$$

have a solution? Well, the highest common factor of 6 and 15 is 3, and 3 does divide 12, so there is a solution. (For example, $x = 7$, $y = -2$ solves the equation.)

Notice that for a given Diophantine equation, determining whether or not a solution exists is not the same as finding a solution. It may be possible to determine the existence of a solution quite easily and yet be extremely difficult to find one. (In contrast, if you know how to find a solution, you will know at once that there is a solution! If you can find something, then that thing must exist, whereas things can exist without being found.) For linear Diophantine equations in two unknowns, not only is there a simple way of determining whether a solution exists, but there is a mechanical procedure for finding a solution if there is one. Complete details can be found in most elementary textbooks on number theory.* The key to the solution is the *Euclidean algorithm* for the determination of highest common factors, described next.

Given two numbers x and y, let x mod y denote the remainder that is left when x is divided by y (as in chapter 1). To calculate the highest common factor of two given numbers a and b, with a greater than b,

---

*For example, see *Elementary Number Theory*, by David Burton (Allyn & Bacon, 1980).

proceed as follows: Let a mod $b = r_1$, then let b mod $r_1 = r_2$, then let r_1 mod $r_2 = r_3$. Continue in this way until you obtain a remainder of zero: r_{n-1} mod $r_n = 0$. Then r_n is the highest common factor of a and b.

For example, to find the highest common factor of 133 and 56, proceed as follows:

$$133 \text{ mod } 56 = 21,$$

$$56 \text{ mod } 21 = 14,$$

$$21 \text{ mod } 14 = 7,$$

$$14 \text{ mod } 7 = 0.$$

The highest common factor of 133 and 56 is 7, the last nonzero remainder. (At this stage, you might like to verify for yourself that the numbers 81 and 25 have highest common factor 1.)

The procedure just described appeared in book 7 of Euclid's *Elements*, written around 350–300 B.C., which explains why it is nowadays known as the Euclidean algorithm. But what exactly is an "algorithm"? This question is crucial to Hilbert's tenth problem. Before attempting to answer it, let us look briefly at what else was known about the solution of Diophantine equations when Hilbert gave his lecture.

In fact there was (and still is) very little known. Linear equations with more than two variables can be dealt with by an extension of the Euclidean algorithm method for the two-variable approach, just outlined. For second-degree equations with one or two unknowns, such as

$$x^2 - 3x + 4 = 0$$

or

$$3x^2 - 5xy + y^2 = 7,$$

an impressive theory worked out by Gauss provides a procedure for determining whether a given equation has a solution. (This is the famous quadratic reciprocity theory, mentioned on p. 77). But except for occasional special cases in which clever tricks can be performed, that is more or less all that was known. (A particularly important "special case" concerns the Diophantine equations

$$x^n + y^n = z^n,$$

where n is at least 2. The existence of solutions to these equations for n greater than 2 is the famous problem of Fermat's last theorem, described in detail in chapter 10.)

And now what about the notion of an "algorithm"?

Algorithms and Turing Machines

Some time around A.D. 825, a Persian mathematician named al-Khowarizmi wrote a book outlining the rules for performing basic arithmetic using numbers expressed in the Hindu decimal form that we use today (with columns for units, tens, hundreds, and so on, and decimal points to denote fractional quantities). From his name comes the modern word *algorithm*.

An *algorithm* is a step-by-step method for performing some kind of calculation. Exactly how the instructions are written down or specified is not important. What is important is that the instructions be complete and unambiguous, with no room for choice or chance, and should work for all possible starting data, not just particular values. The Euclidean algorithm described in the last section is a good example. The instructions tell you exactly what to do at each step, and the procedure works for any numbers a and b (with a greater than b—to allow for all cases, simply add an initial instruction to write the two numbers with the greater preceding the smaller). Other examples of algorithms are the rules for addition, subtraction, multiplication, and division of numbers written in decimal format that al-Khowarizmi laid down in his book.

Hilbert's tenth problem asks if there is an algorithm to determine whether a given Diophantine equation has a solution. For some particularly simple Diophantine equations, there is. For instance, there are algorithms for linear and quadratic equations in at most two unknowns, as mentioned earlier. But is there an algorithm that works in all cases? If the answer is yes, to prove this it would be enough to write down the appropriate algorithm. But suppose the answer is no. How do you prove that? The notion of a "step-by-step instruction set," though adequate for recognizing a specific algorithm as such, is too vague to enable us to prove that there is no algorithm for performing a certain task. For that, a rigorous, mathematical definition is required.

Given present-day computer technology, we could formulate a definition of "algorithm" in terms of a computer program written in a specific computer language for a specific machine, and this would certainly be precise enough. But this creates problems. First, which language and which machine? And what about limitations on the size of numbers that the machine can handle or the amount of memory that is available? If you are prepared to remove all limitations on data size, the choice of the language and machine will not be important to the resulting definition of "algorithm." All languages and machines will produce precisely the same collection of calculable functions; that is, a calculation will be possible on one machine with one language if, and only if, it is possible on any other machine using any other language. This is intuitively clear when you realize that at their most basic level of operation, all that computers do is manipulate zeros and ones.

There is no need, however, to resort to computer technology to obtain a workable definition of algorithm. Several definitions were formulated by mathematical logicians (notable among them Emil Post, Alonzo Church, Stephen Kleene, Kurt Gödel, and Alan Turing) during the 1930s, long before the advent of the computer age. Some quite different approaches were adopted: an "equational calculus," a calculus of "recursive functions," and various abstract "machines." But in each case, the resulting notion of a "performable calculation" was the same, so as far as defining the notion of an "algorithm" is concerned, you can choose any of these approaches. Why not choose the simplest method, which was worked out by the English logician Alan Mathison Turing?

Turing postulated the existence of an abstract computing "machine," known nowadays as a *Turing machine*. This consists of a read/write *head* through which passes a two-way infinite *tape*, divided into squares (see figure 30). The squares on the tape can either be blank or carry one symbol from a fixed alphabet (0 and 1 are sufficient, but the exact choice of alphabet is not important in the long run). At any one time, the head is in one of a finite number of different *states*. (Two states are sufficient, but the exact number is not important.) The operation of the machine proceeds in a step-by-step fashion, with each step consisting of three distinct actions. At any instant, the head is scanning one tape square. What

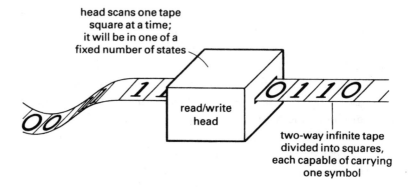

head scans one tape
square at a time;
it will be in one of a
fixed number of states

read/write
head

two-way infinite tape
divided into squares,
each capable of carrying
one symbol

FIGURE 30   A Turing machine. This hypothetical computing device was invented by the English mathematician Alan Turing in the 1930s to provide an abstract framework for studying computation. Despite the simplicity of the Turing machine concept, it can be shown that any computation, no matter how complicated, can be carried out by a Turing machine. The Turing machine concept allows a precise definition of an "algorithm" as a Turing machine program.

The read/write head is at any instant in one of a fixed, finite number of states. The action of the machine proceeds in discrete steps. The actual step at any stage depends on the current state of the head and the contents of the tape square being scanned by the head. The operation of the machine is controlled by a program, which consists of a table indicating what action follows each combination of head state and tape input character (see box B).

it does depends on the contents of that square and the state of the machine. Depending on these two factors, the machine erases the existing tape character and either leaves the square blank or else writes another (or possibly the same) symbol there. Then it moves the tape through the head by one square in either direction, and finally it goes into another (possibly the same) state. The overall behavior of the machine is determined by an *instruction* set that says—for each state and each possible symbol read—exactly which three actions should be performed. The initial data (if there are any) is supplied by writing it on the tape (according to some coding system, the choice of which is not important), and the machine is set in motion, with the head scanning the first data square. If and when the computation ends, the machine enters a special *halt* state and ceases to operate. Any answer that has been produced can then be found on the tape, starting from the square being scanned when the machine stops.

Based on a Turing machine, an *algorithm* consists of a sequence of instructions that determines the behavior of the machine in the manner just indicated. Obviously, with such a simple machine, even the most basic calculation will require a painstakingly detailed "algorithm" (see box B). But the point of the concept is that it provides a precise definition of an algorithm (and of a *computation*) that is simple enough to be handled mathematically and yet is adequate for performing any "algorithmic calculation." This does not mean, though, that such a device should actually be built—although numerous enthusiasts have done just that.

Computable Sets

A crucial notion for solving Hilbert's tenth problem is that of a *computable set* of integers. This is a set S of integers for which there is a mechanical (or algorithmic) method of determining which numbers are in S and which are not. In terms of Turing machines, a set S of integers is said to be *computable* if there is a Turing machine program that, given any integer as input, halts with an output of 1 if the integer is a member of S and halts with an output of 0 if the integer is not a member of S. For example, the set S of even integers is computable, and the program outlined in box B performs the necessary calculation. (That program, however, deals only with positive integers. To allow for negative integers, you need a convention for coding positive and negative numbers, such as using the first symbol as a sign. As an exercise, you might like to try to modify the program given in box B to handle this more general case.)

Notice that in the preceding definition of a computable set, the Turing machine program produces an answer for every input. It cannot, as so often occurs with real programs running on real computers, go into an infinitely recurring loop or start a never-ending search for a data item that does not exist. A weaker notion, which allows for such "unending calculations," is that of a *listable set* of integers (known to mathematicians as a *recursively enumerable set*). In terms of Turing machines, a set S of integers is said to be *listable* if there is a Turing machine program

Box B

A Simple Turing Machine Program

For this example, the alphabet of tape symbols consists of zeros and ones only. Positive integers are represented by a consecutive sequence of ones, with n ones denoting the number n (so the positive integers are denoted by 1, 11, 111, 1111, . . .). The five states are labeled 1, 2, 3, 4, and H (the special halt state). The object of the program is to decide whether a given integer on the input tape is even or odd. If it is even, the machine should output a 1 and stop; if it is odd, it should output a 0 and stop. This output is to appear on the tape after the integer, with one blank square in between. It is assumed that the input integer is aligned with the head initially scanning its first digit and the rest of the integers to the right.

In the following program table, b denotes the blank tape "symbol," and R means scan the next tape square to the right. More complicated programs would probably involve some movements to the left as well.

Condition		Action		
State	*Input*	*Output*	*State*	*Motion*
1	1	1	2	R
1	b	b	3	R
2	1	1	1	R
2	b	b	4	R
3	—	1	H	—
4	—	0	H	—

continued opposite

The action of this program for input 3 (i.e., 111 on the tape) is followed next, step by step. The arrow indicates the tape square being scanned, and the current head state is given.

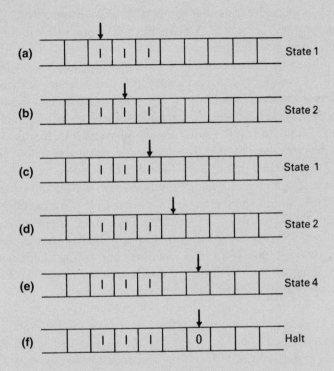

You might like to follow the action of this program for one or two other inputs or work out programs for performing other simple tasks such as the addition of two positive integers represented in this manner.

that, given any integer as input, halts with an output of 1 if, and only if, the integer belongs to S. If the input integer does not belong to S, the program may halt with output 0, or it may not halt at all. Thus if you run the program on a given input integer N and N is a member of S, eventually the program will tell you so, but if N is not a member of S, you might never find this out—the calculation might go on forever, although you could never be sure that it was not just about to stop. So this is a very one-sided situation.

As you might imagine, the two notions of a computable set and a listable set are closely related. A set S of integers is computable if, and only if, both S and $\bar{S}$ are listable, where $\bar{S}$ is the *complement* of S (i.e., the set of all integers that are not members of S). It is fairly easy to see why this is so. If S is computable, any program P that demonstrates this fact will also demonstrate the listability of S. To obtain a program that shows that $\bar{S}$ is listable, add a final step to program P that replaces an output of 0 by 1 and an output of 1 by 0. Conversely, if both S and $\bar{S}$ are listable, to obtain a program showing that S is computable, proceed as follows: Let P and Q be programs that give the listability of S and $\bar{S}$, respectively. If you now feed a given integer N simultaneously into two Turing machines, one running program P and the other Q, if N is a member of S, program P will eventually halt with output 1, and if N is a member of $\bar{S}$, program Q will eventually halt with output 1. Taken together, the two Turing machines give you a mechanical way of deciding whether or not a given integer is in S. Intuitively, this tells you that S is computable. To make it precise in terms of the definitions, you must construct just one Turing machine program that does the job of the two programs P and Q. One way is to write a program R that, for a given integer input, alternately runs P and Q on that input (say 100 steps of each) until one of these halts with output 1. If it is the P part that does this, then R will output 1 and halt; if it is the Q part, then R will output 0 and halt. Thus program R will demonstrate the computability of S.

This result is the best that can be achieved in establishing the relationship between computability and listability. It is definitely not the case that the two notions are the same. Every computable set is, of course, listable, but some listable sets are not computable. To construct an example of such a set, you need Turing's concept of a "universal" Turing machine—a single Turing machine program that can simulate the operation of every Turing machine program. Turing showed that such a program can be constructed, and although the exact details of the construction are rather technical, the underlying idea is simple. Start with a list of all Turing machine programs: $P_1, P_2, P_3, \ldots$. The universal program runs like this: On the input tape you first put a natural number N. When the universal program reads N, it then runs program P_N (any necessary data being found on the tape following N).

Next, to construct a set S that is listable but not computable, let S be the set of all those natural numbers N such that program P_N halts with output 1 when it runs on input N (where $P_1, P_2, P_3, \ldots$ is the list of all Turing machine programs used to obtain the universal Turing machine program). Using the universal Turing machine program, it is easy to show that S is listable: simply add to the beginning of the universal program a small program that replaces any input integer on the tape with two identical copies of itself, leaving the head scanning the start of the first copy. (It is a good exercise to check for yourself how this works.)

To show that S is not computable, suppose that on the contrary, it is computable. Then, as we just observed, $\bar{S}$ will be listable, so there will be a program that demonstrates this fact. This program must appear in the list, $P_1, P_2, P_3, \ldots$ of all programs, say as P_k. Now either k will be a member of S or it will not. If k is a member of S, then k is not a member of $\bar{S}$, so because P_k "lists" $\bar{S}$, P_k cannot halt with output 1 from k. So k does not satisfy the condition that defines S, and it cannot be a member of S. Thus if k is in S, then it is not! But what happens if k is not in S? Then it is a member of $\bar{S}$, so P_k will halt with output 1 on k, and k will satisfy the defining requirement of S, which means that k is a member of S. That is, if k is not a member of S, then it is! (Does this seem familiar to you? See chapter 2, in particular Russell's paradox and the proof of Cantor's theorem.) Having thus arrived at a contradiction, the inescapable conclusion is that S cannot in fact be computable, as was first supposed.

With this last result available to us, we now can describe the solution to Hilbert's tenth problem.

Hilbert's Tenth Problem

Hilbert did not, in fact, ask directly if there was an algorithm to determine whether a given Diophantine equation has a solution. Rather, he asked for such an algorithm to be produced. To quote his own words: "Given a Diophantine equation with any number of unknown quantities and with rational integral numerical coefficients: to devise a process according to which it can be determined by a finite number of operations whether the equation is solvable in rational integers."

Elsewhere, however, in his lecture he says (about problems in general), "Occasionally it happens that we seek the solution under insufficient hypotheses or in an incorrect sense, and for this reason do not succeed. The problem then arises: to show the impossibility of the solution under the given hypotheses, or in the sense contemplated."

In the case of the tenth problem, this is exactly what happened. What Matyasevich proved in 1970 was that no such algorithm exists.

The first serious attempt to obtain such a result was made by Martin Davis in 1950. His strategy was this (see the example that follows if you find this confusing): prove that for every listable set S, there is a corresponding polynomial $P_S(x, y_1, y_2, \ldots, y_n)$, with integer coefficients, such that a positive integer k belongs to S if, and only if, the Diophantine equation

$$P_S(k, y_1, y_2, \ldots, y_n) = 0$$

has a solution. (The degree of P_S is not important, nor is the number of variables involved. When the problem was finally solved along the lines Davis had suggested, it turned out that P_S need not have degree greater than 4 and n need not be greater than 14.)

For example, let S be the set of all positive integers that are not of the form $4k + 2$, for some k. Thus

$$S = \{1, 3, 4, 5, 7, 8, 9, 11, \ldots\}.$$

In fact, S is exactly the set of numbers that can be expressed as the difference of two squares (i.e., numbers of the form $a^2 - b^2$ for numbers a and b). Thus,

$$1 = 1^2 - 0^2, 3 = 2^2 - 1^2,$$
$$4 = 2^2 - 0^2, 5 = 3^2 - 2^2,$$

but there are no numbers a and b for which

$$6 = a^2 - b^2.$$

The general proof goes like this. If n is in S, it must have one of the forms $4k$, $4k + 1$, or $4k + 3$. In the first case,

$$n = \left(\frac{n}{4} + 1\right)^2 - \left(\frac{n}{4} - 1\right)^2,$$

and in the other two cases,

$$n = \left(\frac{n + 1}{2}\right)^2 - \left(\frac{n - 1}{2}\right)^2.$$

On the other hand, every square is either a multiple of 4 or one more than a multiple of 4, depending on whether it is the square of an even number or an odd number, so the difference of two squares can never be two more than a multiple of 4, and hence numbers not in S are not a difference of two squares.

Suppose now that with the set S (which is obviously listable), we associate the polynomial

$$P_S(x, y_1, y_2) = y_1^2 - y_2^2 - x.$$

Then, as you may verify for yourself, a positive integer k will be a member of S if, and only if, the Diophantine equation

$$P_S(k, y_1, y_2) = 0$$

has a solution, that is, if, and only if, there is an integer solution to the equation

$$y_1^2 - y_2^2 - k = 0.$$

The preceding example works because of the special property of S mentioned. What Davis set out to do was associate an appropriate polynomial P_S with every listable set S. To see why this would imply the nonexistence of the kind of algorithm that Hilbert asked for, suppose, on the contrary, that there was such an algorithm. Let S be a listable but noncomputable set of integers, as constructed in the previous section.

Based on our supposition that an algorithm exists to determine the solubility of Diophantine equations, there is a Turing machine program H that halts with output 1 on input k if the Diophantine equation

$$P_S(k, y_1, y_2, \ldots y_n) = 0$$

has a solution and halts with output 0 on input k if it does not have a solution. But because of the relationship between S and P_S, H now shows that S is computable, contrary to the choice of S. Hence no such program H can exist. In other words, there cannot be an algorithm of the kind that Hilbert requested.

Unfortunately, although the strategy works in principle, Davis was unable to prove that such a polynomial P_S always exists. The key to the problem turned out to be in work begun by Julia Robinson. She investigated the kinds of sets that can be defined by Diophantine equations, and she was able to develop various techniques for handling equations whose solutions increased in an exponential fashion. In 1960, she collaborated with Davis and Hilary Putnam to show that if *just one* Diophantine equation could be found whose solutions behaved exponentially in an appropriate sense, then it would be possible to describe every listable set by means of a Diophantine equation in the manner sought by Davis and, hence, to solve Hilbert's tenth problem. But these three researchers were unable to find such an equation, and it was not until ten years later that Matyasevich succeeded where the three Americans had failed. He did so by using a famous sequence of numbers connected with a twelfth-century problem concerning rabbits.

Fibonacci's Rabbits and Matyasevich's Solution

In 1202, the Italian mathematician Leonardo of Pisa published (under the name Fibonacci, from the Latin *filius Bonacci,* meaning "son of Bonacci") his book *Liber Abaci,* an influential work that introduced the Hindu-Arabic decimal number system to western Europe. Among the problems he considers is the following:

A man puts a pair of rabbits in a certain place surrounded by a wall.
How many pairs of rabbits can be produced from that pair in a year if

every month each pair of rabbits bears a new pair that from the second month on becomes productive?

If we assume that one month elapses before the initial pair starts to produce, that there are no deaths in the rabbit colony, and that each pair continues to produce regularly, it will not take long to see that the number of adult rabbit pairs present month by month is given by the numbers in the sequence

$$1, 1, 2, 3, 5, 8, 13, 21, 34, \ldots ,$$

which is generated by the simple rule that every number after the first two 1s is obtained by adding together the previous two numbers. Thus, $2 = 1 + 1, 3 = 1 + 2, 5 = 2 + 3, 8 = 3 + 5$, and so on.

This simple sequence turns out to have several interesting properties in its own right, as well as some surprising applications. (For instance, it arises in the theory of computer databases and also in the study of the computational efficiency of the Euclidean algorithm.) For Hilbert's tenth problem, the importance of the Fibonacci sequence lies in the fact that it exhibits exponential growth. The nth number in the sequence is approximately equal to

$$\frac{1}{\sqrt{5}} \left[\frac{1}{2} (1 + \sqrt{5}) \right]^n .$$

(The approximation improves as n increases.) This means that based on the Davis-Robinson-Putnam result mentioned earlier, to solve Hilbert's tenth problem, it was sufficient to find a Diophantine equation whose solutions were appropriately related to the Fibonacci numbers. This is exactly what Matyasevich did. To obtain the Diophantine equation that he discovered, first start with the following ten polynomial equations:

$$u + w - v - 2 = 0,$$
$$l - 2v - 2a - 1 = 0,$$
$$l^2 - lz - z^2 - 1 = 0,$$
$$g - bl^2 = 0,$$
$$g^2 - gh - h^2 - 1 = 0,$$

$$m - c(2h + g) - 3 = 0,$$
$$m - fl - 2 = 0,$$
$$x^2 - mxy + y^2 - 1 = 0,$$
$$(d - 1)l + u - x - 1 = 0,$$
$$x - v - (2h + g)(l - 1) = 0.$$

In these equations, the values of u and v are related in such a way that v is the $2u$th Fibonacci number, which is enough to satisfy the requirements of the Davis-Robinson-Putnam result. If you now simply square each of these ten equations and add them all together to obtain one big equation, you will get the desired single equation that resolves Hilbert's tenth problem. You also will get a great deal more.

Looked at in terms of Hilbert's question, the Davis-Robinson-Putnam-Matyasevich solution is a negative one because it demonstrates that there is no suitable algorithm. But in reality, it is a very positive mathematical result. It says that every listable set of integers can be described using a Diophantine equation: if S is a listable set, there will be a polynomial $P(x, y_1, y_2, \ldots, y_n)$, with integer coefficients, such that a number k belongs to S if, and only if, the Diophantine equation

$$P(k, y_1, y_2, \ldots, y_n) = 0$$

has a solution.

For instance, the set of primes is a listable set. (In fact, it is a computable set. It is easy to write a computer program that tests primality—although, as indicated in chapter 1, it is not so easy to write one that does this efficiently.) Consequently, the set of primes may be described by means of a Diophantine equation. Using a little algebraic manipulation, it follows that there is a polynomial $P(x_1, \ldots, x_n)$ whose positive values (as the variables $x_1, \ldots, x_n$ range over all integers) are precisely the primes. This resolves a long-standing question about whether the primes could be obtained as the values of a polynomial function. (Note, however, that not all values of the function are primes—the function also produces negative values, which may or may not be negatives of primes. But the positive values range over all the primes, and no other positive values occur.)

Unfortunately, the Matyasevich result merely implies the existence of such a prime-generating polynomial. It does not indicate how to construct one, and it took a considerable effort before James Jones, Daihachiro Sato, Hideo Wada, and Douglas Wiens finally found one in 1977. Their polynomial has 26 variables and is of degree 25. It is

$$
\begin{aligned}
(k+2) \{ &1 - [wz + h + j - q]^2 \\
&- [(gk + 2g + k + 1)(h + j) + h - z]^2 \\
&- [2n + p + q + z - e]^2 \\
&- [16(k+1)^3(k+2)(n+1)^2 + 1 - f^2]^2 \\
&- [e^3(e+2)(a+1)^2 + 1 - o^2]^2 \\
&- [(a^2 - 1)y^2 + 1 - x^2]^2 - [16r^2y^4(a^2 - 1) + 1 - u^2]^2 \\
&- [((a + u^2(u^2 - a))^2 - 1)(n + 4dy)^2 + 1 - (x + cu)^2]^2 \\
&- [n + l + v - y]^2 - [(a^2 - 1)l^2 + 1 - m^2]^2 \\
&- [ai + k + 1 - l - i]^2 \\
&- [p + l(a - n - 1) + b(2an + 2a - n^2 - 2n - 2) - m]^2 \\
&- [q + y(a - p - 1) + s(2ap + 2a - p^2\, 2p - 2) - x]^2 \\
&- [z + pl(a - p) + t(2ap - p^2 - 1) - pm]^2 \}.
\end{aligned}
$$

(Note the apparent paradox that the formula appears to split into two factors. What happens is that the formula produces only positive values when the factor $k + 2$ is prime and the second factor is equal to 1.) A suitably positive result with which to end this chapter!

Suggested Further Reading

Both Hilbert's famous 1900 paper to the International Congress of Mathematicians and the solution to Hilbert's tenth problem can be found in the book *Mathematical Developments Arising from Hilbert Problems*, vol. 28 of the series Proceedings of Symposia in Pure Mathematics, edited by Felix Browder (American Mathematical Society, 1974).

7

The Four-Color Problem

Computer Mathematics Comes of Age

In 1976, Kenneth Appel and Wolfgang Haken, two mathematicians at the University of Illinois, announced that they had solved a century-old problem concerning the coloring of maps. They had, they said, proved the *four-color conjecture.* This in itself was a newsworthy event. At that time, the four-color problem was, after Fermat's last theorem (see chapter 10), probably the second most famous unsolved problem in mathematics. But for mathematicians, the most dramatic aspect of the whole affair was the way that Appel and Haken had achieved their proof. Large and crucial parts of their argument were carried out by a computer, using ideas that themselves had been formulated as a result of computer-based evidence. So great was the amount of computing required that it was not feasible for a human mathematician to check every step, which meant that the whole concept of a "mathematical proof" had suddenly changed. What had been threatening to occur ever since electronic computers were first developed in the early 1950s had finally happened; the computer had taken over from the human mathematician part of the construction of a real mathematical proof.

Until then, a *proof* had been a logically sound piece of reasoning by which one mathematician could convince another of the truth of some assertion. By reading a proof, mathematicians could convince themselves of the truth of the statement in question and also understand the reasons for its truth. Indeed, a proof worked as a proof precisely because it did provide those reasons.

Very long proofs such as the classification theorem for simple groups (described in chapter 5) tend to stretch this simplistic view of a proof to a considerable extent, since the average mathematician, faced with a proof that occupies (say) two 500-page volumes, would be tempted to skip over many of the details. But this is really only a matter of economy of effort. Secure in the knowledge that others have checked the various parts of the argument, the busy mathematician need not examine every step in detail. Such proofs are still the product of human endeavor alone. Although computers were used in proving parts of the classification theorem for simple groups, all the results they produced could be checked by hand. The role played by the computer was in no way an "essential" one.

In the proof of the four-color conjecture, however, the use of the computer was absolutely essential—the proof hinged directly on it. To accept the proof, you first must believe that the computer program used does what its authors claim it does. When Appel and Haken submitted their proof for publication in the *Illinois Journal of Mathematics,* its editors arranged for the computer part of the proof to be checked by running an independently produced computer program on another machine! Thus, a critical part of the proof remained hidden from human eyes.

At first, many mathematicians were skeptical. "Such a procedure, which makes essential use of the results obtained on a computer, results that by their very nature cannot be checked by human hand, cannot be regarded as a mathematical proof," argued one critic. For such people, the four-color problem remained unsolved. And indeed, the question of whether or not a "standard" proof can be found remains open to this day. Given the complexity of the computations involved, even supporters of computer-aided proofs must concede that the opponents' views have some justification, and as long as ten years after the proof was first announced, rumors still surfaced periodically that a subtle error had been found in the computer program, thereby making the proof useless. But by and large, with the passage of time and the growing use of computers in society, the number of mathematicians who refused to accept the proof of the four-color theorem gradually declined, and these days almost all mathematicians acknowledge that the arrival of the computer has changed not only the way much mathematical research is carried out but also the very concept of what is regarded as a proof. Checking the program that produces the "proof" now must be allowed as a valid mathematical argument.

And so what is this problem whose solution had such a profound effect on the very nature of mathematics? The story begins almost exactly one hundred years before the first commercial computers were built.

Guthrie's Problem

One day in October 1852, shortly after he had completed his studies at University College, London, the young mathematician Francis Guthrie (who eventually became a professor of mathematics in South Africa) was coloring in a map showing the counties of England. As he did so, it occurred to him that in order to color any map (drawn on a plane) subject to the obviously desirable requirement that no two regions (countries, counties, or whatever) sharing a length of boundary should be given the same color, the maximum number of colors required seemed likely to be four (see figure 31). Being unable to prove this, he communicated the problem to his brother Frederick, still at University College as a student of physics. Frederick passed it on to his mathematics professor, the great English mathematician Augustus De Morgan.

Like Francis Guthrie, De Morgan had no difficulty proving that at least four colors were necessary (i.e., three colors are not sufficient for some maps—see figure 32). He also proved that it is not possible for five countries to be situated so that each of them is adjacent to the other four, which at first glance might appear to imply that four colors always are sufficient but which does not in fact imply this at all (see figure 33), as De Morgan himself appears to have realized. (Many of the numerous false "proofs" of the four-color conjecture that appeared between its formulation in 1852 and its eventual proof in 1976 were based on a belief in this invalid implication. Indeed, it seems that Francis Guthrie himself at one time fell into this trap.)

When he was unable to solve the problem, De Morgan passed it on to his students and to other mathematicians (among them Sir William Hamilton of Trinity College, Dublin, the inventor of quaternions—see

FIGURE 31   A map of the United States. By using four colors, it is possible to color in all the states so that no two states that share a border are colored the same. For example, Colorado and New Mexico must be colored differently, although Colorado and Arizona may be colored the same, since they touch only at a point. In a mathematical study, states such as Michigan, which consist of two physically separate regions, must be regarded as separate entities. You should have no difficulty convincing yourself that it does not seem possible to color the map (in the manner specified) using only three colors.

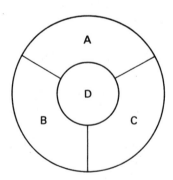

FIGURE 32   Three colors are not enough. To color this map in such a way that no two adjacent countries have the same color, you must color the countries A, B, C, and D using four different colors.

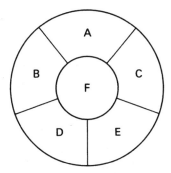

FIGURE 33 Highlighting a false argument. Many people have assumed that because no map allows the configuration in which each of five countries shares a border with the other four, four colors will suffice to color the map. This implication is not valid. The map shown has no configuration in which each of four countries shares a border with the three adjoining ones, and yet the map as a whole cannot be colored using three colors, so the number of colors needed to color a map is not the same as the highest number of mutually adjacent countries in the map.

chapter 3), giving credit to Guthrie for raising the question. But the problem did not seem to have aroused much interest until, on June 13, 1878, the English mathematician Arthur Cayley asked the assembled members of the London Mathematical Society if they knew of a proof of the conjecture. With this act, the hunt began.

Maps, Networks, and Topology

The first major difficulty facing anyone who sets out to prove the four-color conjecture is that it refers to all maps—not just all the maps in all the atlases around the world, but all conceivable maps, maps with millions (and more) of countries of all shapes and sizes. Knowing that you can color some particular map using four colors does not help you at all. You need to produce an argument that will work in all cases, which means that you have very little to go on. In fact, just what do you have to go on? At this stage, it is wise to make sure that we are certain what the problem involves.

For the purposes of Guthrie's problem, a *map* consists of an arbitrary number of regions of the plane ("countries" if you like) separated from one another by lines (or "borders"). This general definition includes maps of the real world, as in figure 31, as well as artificial, "mathematical" maps as in figures 32, 33, and 34. Actually, there is a potential prob-

FIGURE 34   A complicated map that would be difficult (though not impossible) to color using just four colors. (Try it!) This particular map was published as part of an April Fool's joke in the magazine *Scientific American* on April 1, 1975. It was included in an article by the celebrated mathematical columnist Martin Gardner who, with his tongue placed firmly in his cheek, claimed that it was an example of a map that refuted the long-standing four-color conjecture.

lem with the map of the United States (figure 31) in that some states occupy two distinct regions. For instance, Michigan consists of two separate regions on the map, separated by Lake Michigan. Because they are physically separate regions, they must be regarded as such as far as map colorings are concerned. Likewise, Long Island, New York, must be considered a separate entity from the rest of New York State. Thus for a mathematical study of maps, geometry is the dominant notion, not politics.

The four-color conjecture concerns the coloring of the (geographical, not political) regions of a map in such a way that no two regions that share a boundary have the same color. (Regions that just touch at a point, such as Arizona and Colorado in figure 31, are not regarded as having a common boundary and hence may be given the same color.) What is at issue is the smallest number of colors that you need in order to color all the regions of the map in this manner. This is where the second major difficulty lies. Even for a specific map, there are an enormous number of different ways of coloring it, and it is not the number of colors in any one particular coloring that is important, but the least number of colors for which some coloring is possible.

If you think about it for a moment, you will realize that the actual shapes and sizes of the various regions of the map are not important to the coloring problem—only their positions relative to one another. Thus all the maps shown in figure 35 are equivalent for the would-be map colorer. The mathematician would express this by saying that what is at issue is the topological structure of the map.

Topology is a mathematical subject much like geometry. In geometry, you study properties of objects (or figures) in two, three, or more dimensions ("objects" takes on a highly abstract meaning in four or more dimensions, of course). The same is true for topology; the difference between the two disciplines lies in the type of properties considered. In topology, distance and size are not important, nor is straightness or circularity or angle. In fact, topology ignores nearly all the properties that are the lifeblood of geometry, studying instead those properties of objects (figures) that remain unchanged under continuous transformations—for example, bending, stretching, squashing, or twisting. Two-dimensional topology is sometimes referred to as *rubber-sheet geometry*, since it deals with the properties of figures that would not be changed if the figures were drawn on a "perfectly elastic" rubber sheet that was then stretched and twisted (see figure 36).

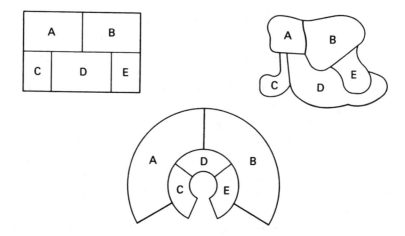

FIGURE 35   Topological equivalence. Each of the maps shown is equivalent in regard to the four-color problem; topologically, there is no difference among any of them.

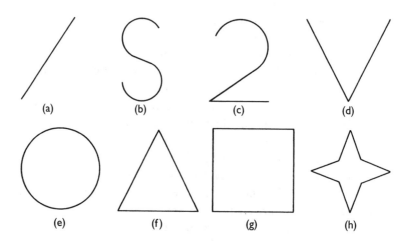

FIGURE 36   Topological properties in two dimensions. The objects (a) to (d) are topologically the same, and the objects (e) to (h) are topologically the same, but none of (a) to (d) is the same as any of (e) to (h).

To anyone encountering the idea of topology for the first time, it would appear that the subject does not contain nearly enough to enable a reasonable mathematical study, but in fact, nothing could be further from the truth. Topology is a vast area of mathematics with many profound results (see chapter 9). Indeed, the four-color problem is itself a problem of topology, although its solution does not use any of the subject's deeper techniques. Figure 35 illustrates why this is so—the shapes and sizes of the countries that make up a map are not important, only their configuration. You will find it helpful in understanding what follows if you keep in mind that it is this *topological* nature of a map that counts, not its superficial "shape."

Once you have grasped this point, the notion of the *neighboring network* of a map will seem to be a sensible alternative way of looking at the four-color problem. Given a map, the neighboring network is obtained as follows (see figure 37). In each region of the map, place a single point, known as a *node* of the network. (If you wish, you can think of these points as the capital cities of the countries.) Then join up the nodes in a certain way to form a network (in much the same way that you might link cities by a rail network). The rule is that two nodes are joined together if, and only if, their respective map regions share a boundary, in which case the line joining them must lie entirely within the two regions, crossing over the common boundary. (For a rail link, this means that the line cannot cross the territory of any third country.)

The neighboring network shows at a glance the topological structure of the map it represents. Indeed, the problem of coloring the map (in the sense of Guthrie's problem) can be reformulated in terms of coloring the network: color the nodes of the network in such a way that any two nodes that are connected have different colors. If all networks can be so colored using four colors, so can all maps, and vice versa. Thus the network formulation of the four-color problem provides an alternative way of looking at it that is equivalent to the original formulation, and it makes sense to investigate such networks.

This brings the problem into the area known as *graph theory*. Notice that as a consequence of the way a neighboring network was defined, no two paths in the network may cross (or intersect). A *graph* is similar to a neighboring network but without this restriction on paths not crossing. (This use of the word *graph* is not connected with its other use in

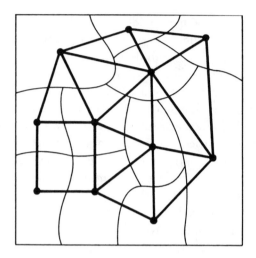

FIGURE 37   Neighboring networks. To obtain the neighboring network for a given map, place a point inside each region of the map and connect these "nodes" by lines that lie entirely within the two regions concerned. This is possible only when the two nodes lie in regions that share a border, in which case the connecting line will cross that border. The connections therefore reflect the existence of common borders. Coloring the map so that no two adjacent countries are colored the same is equivalent to coloring the nodes of the neighboring network so that no two nodes joined by a path of the network are colored the same. In this example, it is possible to join each pair of nodes with a straight line, but this is not always the case, and curved connections are permitted. (Straightness and curving are not topological properties.)

mathematics to refer to curves representing equations drawn on "graph paper.") Although much of the original impetus for the subject of graph theory was provided by the four-color problem, the study of arbitrary "graphs" is now a large and thriving subject in its own right.

Euler's Formula

A particularly useful observation concerning networks was made (in a different context) by Euler. First a little terminology (whose origin will be explained presently). For a start, let us agree that we shall consider only networks that have the property that it is possible to get from any one node to any other by following a route consisting entirely of the network's path connections. This excludes "pathological" examples of "networks" such as a set of nodes with no connecting paths but includes all the networks we shall need for our study of map coloring. Any such network divides the part of the plane that it occupies into a number of regions that are called *faces*. The nodes of the network are sometimes (and particularly in connection with Euler's formula) called *vertices* of the network. The paths that connect these vertices are called *edges*.

At this stage, you should draw a number of networks, and for each one, tabulate the number (V) of vertices, the number (E) of edges, and the number (F) of faces, as in figure 38. When you have done this, work out the quantity $V - E + F$ in each case. You will find that the result is always 1. The fact that the equation

$$V - E + F = 1$$

holds for any network was first proposed by Euler himself.

In fact, Euler was concerned not so much with networks as with polyhedra, which explains the use of the words *vertex, edge,* and *face.* For any polyhedron, you will find that

$$V - E + F = 2$$

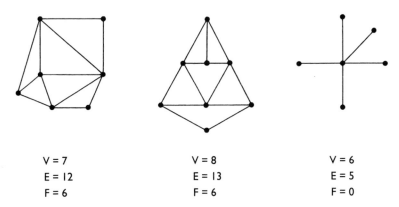

V = 7	V = 8	V = 6
E = 12	E = 13	E = 5
F = 6	F = 6	F = 0

FIGURE 38 Euler's formula. For any network, the number (V) of vertices, the number (E) of edges, and the number (F) of faces are such that $V - E + F = 1$.

(where V, E, and F have their obvious meanings in terms of a polyhedron). To see that this is essentially the same result as the one stated a moment ago for networks, notice that if you remove one face from a polyhedron and then "stretch out" the remaining figure to lie in a plane, the former edges of the polyhedron will form a network connecting the former vertices (the nodes of the new network), and conversely, you can "bend around" any network into the shape of a polyhedron with one face missing. It is the removed or missing face of the polyhedron that accounts for the difference between the formula for the network and the one for the polyhedron.

The proof of Euler's network formula is an excellent example of the kind of argument used in both graph theory and the study of the four-color problem. Suppose you wish to prove for a given network that $V - E + F = 1$. What will happen if you remove an outside edge of the network (assuming there is one)? Both E and F will decrease by 1, but V will remain the same (see figure 39). So the quantity $V - E + F$ remains unaltered by this action. Again, what will happen if the network has a "dangling" vertex (see figure 40) and you remove both the vertex and the edge leading to it? Both V and E will decrease by 1, but F will be unaltered, so in this case too the quantity $V - E + F$ is unchanged. Now, if you start with your given network and, like the sea eroding an island, whenever possible you remove the outside edges and dangling vertices, you will eventually end up with only a single vertex. That is, you will have

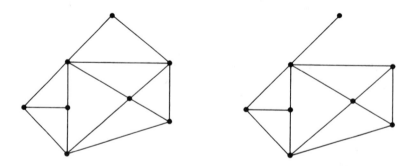

FIGURE 39   Removing an outside edge from a network decreases both E and F by 1 but leaves V unaltered. This does not affect the value of the quantity $V - E + F$.

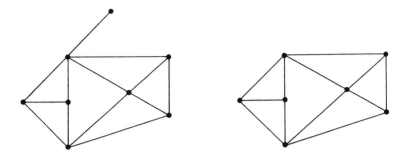

FIGURE 40   Removing a "dangling" vertex from a network decreases both V and E by 1 but leaves F unaltered. This does not affect the value of the quantity $V - E + F$.

reduced your original network to a trivial one in which $V = 1$, $E = 0$, and $F = 0$. In this final "network," the quantity $V - E + F$ is equal to 1. But none of the "erosion" operations you performed altered $V - E + F$. So the value of this expression *in the original network* must have been 1 as well. And that proves the result. If you wish, you can try this out for yourself. From some arbitrarily drawn network, keep removing outside edges and dangling vertices, tabulating the values of V, E, F, and $V - E + F$ as you go.

De Morgan's Theorem

The one positive result that De Morgan appears to have obtained on Guthrie's problem was to prove that in no map can there be five countries such that each borders the other four. By using the neighboring network together with Euler's formula, this is quite easy to demonstrate. For networks, what De Morgan's result amounts to is that it is not possible to draw a network with five vertices so that each vertex is connected to the other four. Certainly, if you try to do this (see figure 41), you will invariably find that you are left with two vertices that cannot be connected without crossing a path that has already been drawn, but this does not constitute a proof, since you may just have drawn the earlier connections inappropriately. The following argument does not depend on a particular drawing and thus provides a rigorous proof.

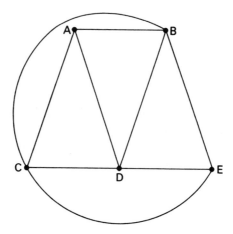

FIGURE 41   It is not possible to draw a network with five vertices so that each vertex is connected to the other four. No matter how you try to join the vertices, you will be left with two vertices (A and E in the network shown) that cannot be joined without crossing one of the lines already drawn.

Suppose that it were possible to draw a network with five vertices such that each vertex is connected to all the others. If the area surrounding the network is regarded as an additional "face," then each edge of the network will separate two faces. Moreover, since there is now one extra face, Euler's formula becomes

$$V - E + F = 2.$$

We know the value of V here; it is 5. Also, since every vertex is joined by an edge to every other, $E = 10$. (Check this for yourself.) So F must be 7 if the preceding Euler formula is to hold.

So far so good. Now we shall perform another calculation. Notice that each face is surrounded by at least three edges. (This is true also for the new "face" we introduced a moment ago, although the word *surrounded* is used in a topological sense in this case.) Thus, counting edges by faces gives at least $3 \times 7 = 21$ edges. But if you count edges in this way (by faces), each edge will be counted twice, since it separates two faces. So the correct answer is that there must be at least $\frac{1}{2} \times 21 = 10\frac{1}{2}$ edges, which (since there is no such thing as half an edge) means that there are at least 11 edges. But as we just noted, $E = 10$. So we have arrived at a contradictory situation, and as usual in such arguments, the conclusion is that the original assumption of the argument must be false—that is, it is *not* possible to draw a network with five vertices so that each vertex is connected to all the others, and this proves De Morgan's theorem about maps.

The Five-Color Theorem

In 1879, within a year after Cayley posed the four-color problem to the London Mathematical Society, one of its members, a barrister named Alfred Bray Kempe, published a paper in which he claimed to prove the conjecture. But he was mistaken, and eleven years later Percy John Heawood pointed out a significant error in Bray's argument. Heawood was, however, able to salvage enough of it to prove that five colors are always enough, and the proof of this "five-color theorem" is sufficiently simple to give here.

First, notice that by reasoning as we just did to relate the Euler formula for networks to that for polyhedra, we may conclude that it does not matter (in regard to the four- or the five-color theorem) whether we draw our maps on a plane or on the surface of a sphere. If we start with a map on a sphere, we can convert it to an equivalent map on a plane by piercing a hole in the middle of one of the regions and pulling the entire map out flat (so that the pierced region becomes one that surrounds the rest of the map). Conversely, if we are given a map on the plane, we may regard the region surrounding the map as an extra country and fold the entire map around into the shape of a sphere (bringing the added surrounding region together to form an "enclosed" region just like all the others). This procedure shows that if every planar map can be colored with N colors, so can every map on the sphere, and vice versa.

We shall in fact prove the "five-color theorem" for maps drawn on a sphere. We shall make use of the Euler formula, which for maps on a sphere is

$$V - E + F = 2.$$

Our use of this formula will be in connection with the map itself, rather than with the associated neighboring network, as was the case with the proof of De Morgan's theorem. (Consequently, a *face* is a region of the map, an *edge* is a border, and a *vertex* is a point where three or more borders meet.)

The idea of the proof is to start with a given (entirely arbitrary) map drawn on a sphere and gradually modify it by a process of merging two or more adjacent countries into one, so that eventually we get a map having at most five countries—which obviously can be colored using five or fewer colors. Provided the steps used in the modification process do not reduce the number of colors required to color the map, this will prove that five colors suffice for the original map. The crux of the proof, therefore, is to describe the individual processes used to *reduce* a given map to a simpler one (i.e., one with fewer countries) without reducing the number of colors necessary to color the map. There are six different *reduction processes,* each of which is applicable to a different situation depending on the map's particular configuration of countries.

First, if one region is entirely surrounded by another (see figure 42(a)), the inner region may be merged with the surrounding one. Any coloring of the new map using at least two colors can be extended to a coloring of the original map using the same colors; the inner region is simply assigned a color other than the one used to color the entire merged region on the modified map.

The next reduction operation applies whenever there is a vertex at which more than three regions touch. For if at least four regions touch, then (and you may need to think about this for a moment) one pair of these regions will not have a common border (anywhere on the map) and these two regions can be merged into one (see figure 42(b)). Given any coloring of the modified map, the original map may be colored using the same number of colors by assigning the same color to the two regions that were merged and coloring the rest of the map the same in both cases. By applying this reduction repeatedly, the map can be modified so that only three regions touch at each vertex. This will be assumed to be the case for the remainder of the reductions.

If there is a region that borders on just two others (see figure 42(c)), that region may be "removed" by merging it with one of these two. If the new map can be colored using at least three colors, the original map can be colored using the same colors simply by coloring the merged central region differently from the two surrounding areas.

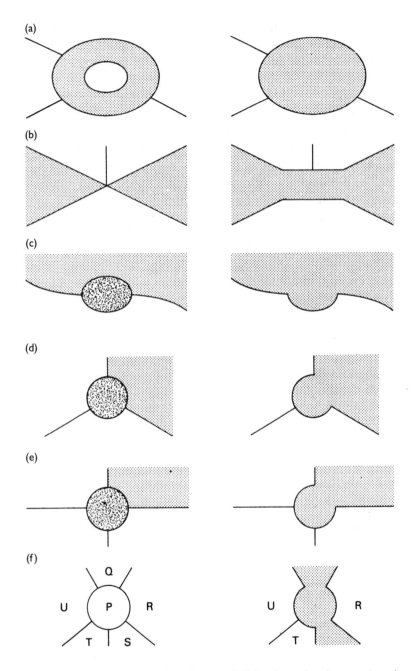

FIGURE 42 Reductions used in the proof of the five-color theorem (see the text for details).

Any region having three neighbors can be "removed" by merging it with one of its neighbors (see figure 42(d)), and as in the previous case, if the new map can be colored using at least four colors, the original can be colored using the same colors.

Likewise, any region having four neighbors can be merged with one of its neighbors (see figure 42(e)), and this will not involve any change in the color requirements when five colors are available.

By applying these reduction procedures as often as possible, you will end up with a map in which no region is surrounded by another, in which each vertex lies on exactly three edges, and all of whose regions have at least five edges. In fact, at least one region will have exactly five edges, as we shall prove next.

There are V vertices, E edges, and F regions. Let a be the average of the number of edges bordering each region. (Note that a might be fractional.) Since each edge lies between two regions,

$$2E = aF.$$

Also, each vertex lies on three edges, and each edge joins two vertices, so

$$3V = 2E.$$

Hence

$$3V = 2E = aF.$$

Substituting $V = \frac{1}{3}aF$ and $E = \frac{1}{2}aF$ in Euler's formula $V - E + F = 2$ gives

$$\tfrac{1}{3}aF - \tfrac{1}{2}aF + F = 2,$$

so

$$a = 6 - \frac{12}{F}.$$

Thus a is less than 6. Since the average number of edges for each region is fewer than 6, some regions must have fewer than 6 edges. But all regions have at least 5 edges. So some regions will have exactly 5 edges, as we set out to prove.

Now consider such a five-edged region P, with neighbors Q, R, S, T, and U, as in figure 42(f). One pair of neighbors of P do not touch, say Q and S. Merge the three regions P, Q, and S. If the new map can be colored with five colors, so can the original. In the merged map, Q and S have the same color, so four colors surround P, which leaves one to spare for P.

This completes the reduction procedure. Since each step reduces the number of regions on the map, by applying it repeatedly you will eventually arrive at a map with five or fewer regions. Since any such map can obviously be colored with five colors, so can the original map. Indeed, by working back through the various reductions, a coloring of the map using five colors can actually be built up in an entirely mechanical way. (Try it out for yourself, starting with a moderately complicated map.)

The basic idea of reducing the map step by step, as in the preceding proof, has been used in almost every serious attempt to prove the four-color conjecture (including the final successful attempt) since Kempe first introduced the device in his flawed attempt to solve the problem. Because the actual solution was essentially an extension of what Kempe did (although a quite considerable extension), we shall take a closer look at his argument.

Kempe's Method

The preceding proof of the five-color theorem, which Heawood salvaged from Kempe's false argument, is much simpler than the one Kempe thought adequate for the four-color theorem. (Clearly, the arguments used in the reductions illustrated in figure 42(e) and (f) do not work when only four colors are available.) Kempe's procedure was roughly as follows:

Assume the existence of a map for which five colors are required (i.e., four colors are definitely not sufficient), then argue to obtain a contradiction. The first step is to define the notion of a *normal map*, that is, one in which no country is entirely surrounded by any other and no more than three countries meet at any point. By applying the first two of the reduction procedures just described, starting from a map that requires five colors, you can obtain a normal map requiring five colors. Since the existence of a map requiring five colors is assumed, it follows that there will be a normal map that requires five colors. There may in fact be many such maps, with different numbers of countries. Among these there will be at least one having the smallest number of countries (for such a map). Kempe tried to obtain his contradiction by working with such a *minimal normal map* requiring five colors.

The point of using a minimal map is that any (normal) map with fewer regions may be colored using four colors, so if you can find a reduction operation that will reduce the size of your map by even one country without altering the requirement for five colors, you will have your contradiction, since the reduced map cannot simultaneously be colorable using four colors and not colorable with fewer than five colors.

Kempe proved (correctly) that in any normal map there must be a country with at most five neighbors; that is, at least one of the configurations shown in figure 42(c), (d), (e), and (f) must occur somewhere in the map. He then argued (incorrectly) that if a minimal normal map requiring five colors has a country with at most five neighbors, it can be reduced to a normal map with fewer countries that still requires five colors, thereby arriving at a contradiction, as outlined a moment ago. His arguments were correct for countries with two, three, or four neighbors. The arguments for the two- and three-neighbor cases were given a moment ago in the proof of the five-color theorem. But if there are four neighbors, we need a different and more subtle argument. It is necessary to examine the part of the map that surrounds the configuration and possibly alter the colors of some of the surrounding countries. It is not unduly difficult, but it does require considerable thought and ingenuity. Where Kempe went wrong was—as Heawood noticed—in the case of a country with five neighbors (see figure 42(f)).

Nevertheless, Kempe's argument already contained two of the key ideas that were eventually used in solving the problem. One is the notion of an *unavoidable set* of configurations—a collection of map configurations such that any minimal normal map requiring five colors must contain at least one of them. (Kempe's unavoidable set consisted of the configurations shown in figure 42(c), (d), (e), and (f).) The other is the notion of *reducibility:* if a certain configuration appears in a minimal normal map requiring five colors, it is possible to reduce the number of countries in the map so as to produce the contradictory situation of a normal map requiring five colors having fewer countries than the minimal one. If you can demonstrate that every configuration in an unavoidable set *is* reducible you can prove the four-color theorem. Kempe's argument failed because his proof of reducibility did not work for one of the four configurations in his unavoidable set. The Appel-Haken proof succeeded by making a detailed analysis of this flawed last case, and this required the discovery of a different unavoidable set. You will begin to get some idea of the difficulty that these researchers faced when you consider that the unavoidable set they eventually constructed contained almost 1500 configurations. How they discovered such a set will be explained shortly. First, what progress was made on the problem in the years between Kempe's false proof and the Appel-Haken solution?

Heawood's Formula

Heawood's work has already been mentioned. After spotting the flaw in Kempe's argument in 1890, Heawood spent the next sixty years working on the problem, and although he did not solve it, he did make progress on the analogous question for maps drawn on surfaces other than planes or spheres. His solution involves the so-called Euler characteristic of a surface. For any surface, say a sphere, a torus, or even a double torus (see figure 43), if you draw a map covering the entire surface, the quantity $V - E + F$ will work out to be the same no matter how you draw the map—just as it does for maps drawn on a sphere (for which the answer is 2). This quantity, which therefore depends not on the map

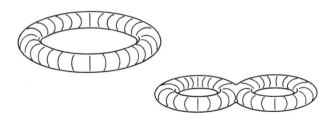

FIGURE 43   A torus and a double torus.

(since any map is as good as any other here) but on the surface (different surfaces give different answers), is called the *Euler characteristic* of the surface. It is a *topological invariant* of the surface, which is to say it remains the same no matter how the surface is topologically manipulated. For the torus the Euler characteristic is 0; for the double-torus it is −2. For the *Klein bottle,** a curious surface that has no edges and only one side and that can be constructed in three-dimensional space only if self-intersection is allowed (see figure 44), the Euler characteristic is 0, the same as for the torus. But note that although they have the same Euler characteristic, the torus and the Klein bottle are not topologically equivalent; you cannot deform one into the other. Although topologically equivalent surfaces do have the same Euler characteristic, topologically different surfaces may or may not.

By employing essentially the same kind of argument used to prove the five-color theorem for maps drawn on a sphere, Heawood proved that for a surface of Euler characteristic *n*,

$$\frac{1}{2}(7 + \sqrt{49 - 24n})$$

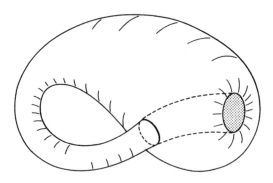

FIGURE 44   The Klein bottle—a topological figure that consists of a surface having no edge and only one side. In three-dimensional space, this can be achieved only if the surface is allowed to "pass through" itself. In four dimensions, such self-intersection would not be necessary.

---

*Named after the German mathematician Felix Klein.

colors suffice to color all maps drawn on the surface, provided n is, at most, equal to 1. Unfortunately, the only surface for which n is greater than 1 is the sphere, for which $n = 2$, and so Heawood's otherwise splendid result just failed to handle the one case that everyone was most interested in—for if you put $n = 2$ in Heawood's formula, you get the answer 4, of course!

Thus for the torus, for which $n = 0$, seven colors are sufficient, and since it is not hard to draw a map on the torus that cannot be colored with six colors, this proves the "seven-color theorem" for maps drawn on a torus in the strong sense that seven colors suffice and fewer colors do not. In fact, in 1968 Gerhard Ringel and J. W. T. Youngs showed that Heawood's formula gives the exact minimal number of colors required in all cases except for the sphere (for which at that time the answer was not known) and the Klein bottle (for which $n = 0$ and the formula gives the answer 7, although another argument shows that only six colors are needed*).

Toward a Four-Color Theorem

After Heawood's work, numerous mathematicians (and an even greater number of amateurs) investigated the four-color problem, developing in the process many mathematical techniques that ultimately proved to have applications elsewhere in mathematics. Of all this effort, some of it can be seen, with the benefit of hindsight, to have led toward the final solution to the problem. Briefly, this is what happened.

In 1913, George Birkhoff improved on Kempe's reduction technique and managed to show that certain configurations larger than Kempe's were reducible. In 1922, Philip Franklin used some of Birkhoff's results to show that every map with 25 or fewer countries could be colored with four colors. In 1926, a chap called Reynolds improved this result to 27 countries, and then in 1938, Franklin regained the "record" by going up to 31. C. E. Winn, in 1940, reached 35, and there progress stopped until 1970, when Oystein Ore and Joel Stemple proved the four-color theorem for all maps having fewer than 40 countries. The figure reached 96 before the Appel-Haken proof finally rendered all such results superfluous.

---

*So, as a consequence of the four-color theorem, it is now known that the Klein bottle is the only surface for which Heawood's formula does not give the exact, minimal answer.

But although all this work did show that many configurations are reducible, the set of all configurations that had been proved reducible by
1970 did not even come close to forming an unavoidable set (as would
be required to prove the four-color conjecture). Various unavoidable sets
had been constructed, but none seemed likely to lead to an unavoidable
set of reducible configurations. You had either reducibility or an unavoidable set, but not both. In 1950, the German mathematician Hein-
rich Heesch, who had been working on the four-color problem since
1936, proved that a finite unavoidable set of reducible configurations
must exist, and he estimated that an unavoidable set of reducible configurations must contain about 10,000 separate configurations. Although
this estimate ultimately proved to be grossly overgenerous, it was accurate in indicating that the problem would be solved only with the aid
of very powerful computational equipment capable of handling vast
amounts of data. Indeed, Heesch himself, realizing that the ability to
handle large sets of configurations was likely to be the key to the solution, was the first to advocate—and attempt—a computer-aided assault
on the problem.

He began by formalizing the several known methods of proving configurations reducible and noted that at least one of them (a straightforward generalization of Kempe's method) was sufficiently mechanical to
be implemented on a computer. Karl Dürre, a student of Heesch, then
wrote a program to prove reducibility. All this was done in terms of the
neighboring network representation of the map, which provided a more
convenient way of handling the problem on a computer.

One problem to be overcome was that the failure of one method of
proving that a particular configuration was reducible did not necessarily mean that the configuration could not be reduced, since another
method might succeed where the first had failed. To deal with this difficulty, it was necessary to develop a small "arsenal" of techniques for
proving reducibility. By the late 1960s, Heesch had built up a large
enough arsenal for Appel and Haken to use at the start of their final assault on the problem in 1976.

However, comparable progress had not been made with the construction of an unavoidable set of configurations. Heesch did try a
method that was analogous to moving a charge around an electrical network, but he did not pursue it very far. Perhaps he should have. For this
was the trick that led to the final solution.

Heesch's Charge Method

The neighboring network associated with a minimal normal map requiring five colors is (and this follows from Kempe's work) one in which every face is a triangle and at least five edges meet at each vertex. (The number of edges meeting at a vertex is called the *degree* of that vertex.) The idea now is to regard the network as an electrical one and to assign a charge to each vertex. If a vertex has degree k, it is given charge $6 - k$. Thus vertices of degree 5 have a positive charge (of $+ 1$); vertices of degree 6 have no charge; vertices of degree 7 have charge -1; and so on. It follows from Kempe's work that the sum of the charges over the entire network is (for any network) 12. The precise value of 12 is not important; what is important is that the sum of the charges is always positive.

Now suppose you start to move positive charge around the network (in fractional amounts if you wish). This will not, of course, cause any net gain or loss in the total charge in the network, but some degree-5 vertices may end up losing all their charge (i.e., become *discharged*), and some vertices of degree greater than 6 may end up with a positive charge (i.e., become *charged*). The exact final situation depends on the precise *redistribution* (or *discharging*) procedure used. But (and this is the key), because it is possible to determine the layout of small pieces of a map without knowing the entire map, then given a specified discharging procedure (applicable to any map), it is possible to generate a finite list of all the configurations that will end up with net positive charge.

Because the total charge on the network is positive, there will always be some vertices with positive charge. So, because all possible receivers of positive charge are included in the finite list of configurations that has been generated from the discharging procedure, every network (of the kind now being considered) must contain at least one of these configurations. In other words, the list of configurations generated will form an unavoidable set, which is what we are looking for. Kempe's original unavoidable set can be regarded as the one resulting from the trivial procedure of moving no charge at all. Thus the discharging procedure is a generalization of Kempe's method. And because it is more general, it should have a greater chance of success.

A simple example should help make things clear (though it is likely that you will need to "play around" with this example for some time in order to see what is going on). Suppose your discharging procedure consists of transferring $\frac{1}{5}$ of a unit of charge from every degree-5 vertex to each neighboring vertex having degree 7 or more. Then the resulting unavoidable set consists of just the two configurations shown in figure 45. To see how these arise, note first that a degree-5 vertex can end up positive only if it has at least one neighbor of degree 5 (figure 45(a)) or degree 6 (figure 45(b)). A degree-6 vertex starts out with no charge and does not receive any under this procedure. A degree-7 vertex can become positive only if it has at least six degree-5 neighbors. If this is the case, then because every face is a triangle, two of these neighbors are joined by an edge (so figure 45(a) applies to that pair of neighbors). A vertex of degree 8 or more cannot become positive even if all its neighbors have degree 5. Moving $\frac{1}{5}$ of a unit simply is not enough. (For example, for a degree-8 vertex with eight degree-5 neighbors, the original charge is -2, and $8 \times \frac{1}{5} = 1\frac{3}{5}$ units are transferred, leaving a final charge of $-\frac{2}{5}$.)

Thus the two configurations shown form an unavoidable set. That

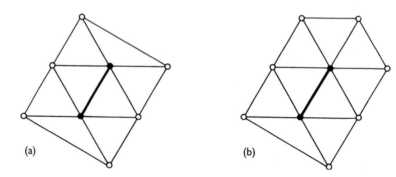

(a)   (b)

FIGURE 45   An unavoidable set generated by the simple discharging procedure described in the text. The unavoidable set consists of the configurations (a) and (b). In (a), the configuration is produced by two degree-5 vertices joined together; in (b), by a degree-5 vertex joined to a degree-6 vertex. These two vertex pairs are shown as black circles connected by a heavy line. The remainder of each of the two configurations is then determined by the degrees of the vertices in the producing pair and the fact that the network is triangular. (There is no restriction on the degree of each of the outer vertices, shown as white circles.) Unavoidability means that at least one of the two networks (a) and (b) is always found.

is, because this argument works for any network (of the type being considered), at least one of these two configurations is always found.

The idea behind using the charge method to prove the four-color conjecture is to find a discharging procedure such that the resulting unavoidable set consists entirely of reducible configurations. If this can be done, the theorem follows immediately. (Neither of the two configurations produced in the example just given is reducible, by the way.)

Proof of the Four-Color Theorem

In 1970, Haken devised some new methods for improving discharging procedures, and although the task seemed to be a formidable one that would require immense amounts of computation time, he began to hope that this might lead eventually to a proof of the four-color conjecture. In 1972, he and Appel began working in earnest to try to bring this hope to fruition.

Their goal was to devise a discharging procedure that would lead to an unavoidable set of reducible configurations. This involved two things: finding the discharging procedure and proving the reducibility of the unavoidable configurations that it produced. At first, they worked with highly restricted types of network that previous work (by Heesch and others) had shown should be easier to deal with. The general strategy was clear: start with a promising-looking discharging procedure, and try to prove that each of the resulting unavoidable configurations is reducible. If reducibility cannot be proved for one or more configurations in the list, amend the discharging procedure so that the troublesome configuration(s) no longer appear. Although simple to describe, carrying out this strategy was not at all easy and involved many weeks of human-machine "dialogue" as first one and then another discharging procedure was tried, but very gradually progress appeared to be made. A similar approach of computer "experiment" punctuated by human intervention was adopted as the pair carried out a simultaneous search for improved methods of proving reducibility. After three years of this, by the beginning of 1976, they finally felt they had enough information to begin their assault on the full problem. As a result of all their experimental work, they had developed a discharging procedure that looked likely to produce an unavoidable set of reducible configurations, and they had written a routine for proving reducibility that looked like it would work for the kinds of configurations they would encounter. Their computer program had a self-modifying feature so that if it encountered a configuration that could not be proved reducible, the program would move around positive charge to try to avoid the difficulty. But would it work? Could it work? The only way to find out was to start the program and see what happened. And that is what they did.

Six months later, in June 1976, they had their answer. Their program had (with considerable help from its two by now very expert creators) succeeded in proving the four-color theorem. It had taken four years of intense work and 1200 hours of computer time. The discharging procedure that finally did the trick differed from the one they had started with by some 500 modifications, suggested by the results of the various computer runs. The two mathematicians themselves had to analyze some 10,000 neighborhoods of positively charged vertices *by hand calculation,* and the computer had to examine more than 2000 configurations and prove the reducibility of a total of 1482 configurations in the unavoidable set. But it all worked. A hundred years of effort had finally come to an end.

Mathematics would never be quite the same again.

Aftermath

Even though the Appel-Haken proof brought an end to the 124-year quest for a solution to the four-color problem, it was certainly not the end of studies of map coloring. For whereas interest in the original problem was generated largely by curiosity, the general issue of map colorability has significant applications. More precisely, the intimately related, and completely equivalent, question of graph colorability is important in a number of application areas. One crucial application is in the area of task scheduling. Suppose the nodes of a graph represent individual tasks in a complex industrial process, with an edge between two nodes indicating that one of the two tasks depends on the completion of the other, so that the two tasks cannot be performed at the same time. Then a coloring of the graph (i.e., a coloring of the nodes so that no two nodes connected by an edge have the same color) splits the tasks into disjoint collections that can be performed simultaneously.

Of course, a graph that arises from an industrial application may not be able to be drawn on a plane (or a sphere), so the question of colorability of graphs that cannot be drawn on a plane is relevant. The technical term for a graph that cannot be drawn on a plane is *nonplanar*. For example, the graph K_5 consisting of five vertices each joined by an edge to the other four is nonplanar. No matter how hard you try, every time you try to draw this graph on a plane, you will be forced to draw one edge that crosses another (see figure 46(a)). Another nonplanar graph is $K_{3,3}$, a graph consisting of two groups of three vertices, with each vertex in one group joined to all (and only) the three vertices in the other group (see figure 46(b)). Neither of these two graphs can be drawn on a plane, although each can be drawn with ease on (say) a torus.

In 1930, the Polish mathematician Kazimir Kuratowski proved that K_5 and $K_{3,3}$ are essentially the only nonplanar graphs, in that any nonplanar must contain at least one of these two graphs as a "minor." A *minor* of a graph is the graph that results when groups of vertices joined by edges coalesce to give a single vertex. For maps, this corresponds to neighboring countries erasing their border and forming a federation.

Clearly, any graph (on any surface) that contains K_n (the analogue of K_5 with n vertices in place of 5) cannot be colored using fewer than n colors. (Because each vertex of a complete graph such as K_n is joined to every other vertex, so it cannot be colored with fewer than n colors.) In

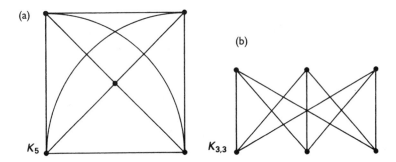

(a)

(b)

K_5

$K_{3,3}$

FIGURE 46   The two graphs K_5 and $K_{3,3}$ are nonplanar. Neither can be drawn on a plane without at least one edge crossing another. Kuratowski proved that any nonplanar graph has at least one of these graphs embedded in it.

1943, Hugo Hadwiger conjectured that any graph that does not contain K_{n+1} as a minor can be n-colored. The conjecture is obviously true for $n = 1$ and $n = 2$ and easy to prove for $n = 3$. It is also true for $n = 4$, but the proof is not easy. In 1992, Neil Robertson, Paul Seymour, and Robin Thomas proved the conjecture for $n = 5$.

That leaves a lot of unresolved cases—indeed, infinitely many. But with important applications waiting in the wings, the increased understanding that results from such advances can be of considerable significance, perhaps far more significance than the four-color theorem that created all the interest in graph coloring in the first place. The four-color theorem might have been put to rest in 1976, but work on the question of map (and graph) colorability continues to this day.

Suggested Further Reading

In response to a number of criticisms that their proof was flawed, Appel and Haken wrote a highly readable account of their proof for the magazine *The Mathematical Intelligencer* 8 (1986):10–20. The title of the article is "The Four Color Proof Suffices," a play on the special postmark that the local Illinois post office used to honor the solution to the problem in 1976, which announced proudly "four colors suffice," no doubt much to the bafflement of the uninitiated who happened to receive a letter bearing this message.

For a general introduction to graph theory, try *Introduction to Graph Theory*, by Robin J. Wilson (Addison Wesley Longman, 1996).

In the *Journal of Combinatorial Theory*, series B, 70, no. 1 (1997):2–44, Neil Robertson and his colleagues present a newer proof of the four-color theorem. They use the same approach as Appel and Haken did, but their proof is shorter. It looks at only 644 unavoidable configurations (down from almost 1500) and uses only 32 discharging rules (down from more than 300).

A nice, comprehensive description of the development of graph theory in general is contained in *Graph Theory 1736–1936*, by N. L. Biggs, E. K. Lloyd, and R. J. Wilson (Oxford University Press, 1976, 1986).

8

Hard Problems About Complex Numbers

A Complex Subject

For many readers, this will be the most difficult chapter in the book. Not because the mathematics is intrinsically any harder than in other chapters, but because of the degree of abstraction involved. True enough, numbers—both natural and complex—form the core of the subject. But the essential task of *complex analysis* (the term *complex function theory* is used almost synonymously) and the closely allied field of *analytic number theory* (the application of the results and techniques of complex analysis to the study of the natural numbers) is to root out and exploit the deep structure and interconnections that lie beneath what—on the evidence of the description of complex numbers given in chapter 3—seems to be a fairly simple notion. To carry out such a study requires some very abstract mathematical techniques that are not familiar to many people outside mathematics itself. And unfortunately, pictures do not help, since the subject is not very visual (unlike topology, the subject of chapter 9, about which equally difficult and unfamiliar ideas can be conveyed, at least in simple cases, using pictures and diagrams). It is an important field, however, and one in which significant advances have in recent years been made, so it should not be ignored. Moreover, if you persevere with this chapter, you will discover that from all the abstraction come some remarkable insights into such familiar concepts as fractions and prime numbers. (I am assuming that you have already read the introductory account of complex numbers given in chapter 3.)

Although the three problems that form the core of this chapter are all highly abstract, you should not think that complex analysis has no applications outside mathematics. Ever since the first work on the subject by Augustin Cauchy in 1825, complex analysis has always had connections with the outside world. Riemann's work on the subject, following closely after Cauchy's, showed how complex-function theory could greatly assist in solving problems in physics, and further work on the so-called integral transforms (such as the ubiquitous Fourier transform) made this connection even more apparent.

The two-dimensional nature of complex numbers (see chapter 3 for a discussion of the complex plane) enables them to be used to study problems in two dimensions in much the same way that problems in one dimension can be handled using real numbers. Because real-life problems in three dimensions that have a symmetrical nature (such as the flow of liquid through a circular pipe) often reduce to a mathematical problem in two dimensions, complex analysis is relevant to both physicists and engineers.

The Russian mathematician N. Y. Joukowski (1847–1921) used complex analysis to specify the shape of an aerofoil (i.e., the cross section of an aircraft wing) and to study the flow pattern around it, thereby revolutionizing aircraft design. Since then, complex-function theory has become important to the description of all forms of fluid flow and to the design of cars and ships. In 1920, scientists at Bell Laboratories in the United States systematically used complex-function theory to design the filters and high-gain amplifiers that made long-distance telephone calls possible. (If you know an electronics engineer, ask him or her about the importance of the Nyquist criterion for the stability of feedback amplifiers. This is a direct application of complex analysis.) Complex analysis is now, in short, an indispensable aid; very little of present-day science and technology is not dependent on complex numbers in one way or another.

So just what is complex analysis? A good first attempt at an answer would be that it is an extension of the methods of the calculus (differentiation, integration, infinite sums, and so on) from the more familiar setting of the real numbers into the realm of the complex numbers. But like all first attempts, this answer is at the same time helpfully suggestive and misleading, because the whole apparatus of the calculus takes on an entirely different form when developed for the complex numbers. Concepts that with the real numbers appear quite distinct from one another may turn out to be closely related when complex numbers are introduced into the picture.

One instance of this, already mentioned in chapter 3, is Euler's identity

$$e^{\pi i} = -1.$$

Another (closely related) example is provided by the equation

$$e^{xi} = \cos x + (\sin x)i$$

(for any real number x), which relates e, i, and the familiar trigonometric functions sine and cosine. In fact, there is nothing to prevent you from applying the sine and cosine functions themselves to complex numbers. Of course, you cannot calculate, say, $\sin (3 + 4i)$ in terms of right-angled triangles, as you can with real numbers. Once you decide to work with complex numbers, you must be prepared to go wherever the theory takes you, and for trigonometric functions, this turns out to be into the realm of *infinite series*. Just as the infinite sum

$$e^x = 1 + \frac{x}{1!} + \frac{x^2}{2!} + \frac{x^3}{3!} + \dots$$

gives a valid answer whether x is real or complex, so too do the infinite-series equations

$$\sin x = x - \frac{x^3}{3!} + \frac{x^5}{5!} - \frac{x^7}{7!} + \dots ,$$

$$\cos x = 1 - \frac{x^2}{2!} + \frac{x^4}{4!} - \frac{x^6}{6!} + \dots .$$

If x is real, each of these infinite sums will give exactly the same answer that you would obtain from the usual geometric definition (provided that you regard the "angle" x as being measured not in degrees but in *radians,* where 1 radian equals $180/\pi$ degrees, that is, about 57.3 degrees).

But there is nothing to stop you using these equations when x is complex.

By simple algebraic manipulation of the preceding three expressions, you can obtain the formulas

$$\sin x = \frac{1}{2i}\,(e^{xi} - e^{-xi}), \quad \cos x = \frac{1}{2}\,(e^{xi} + e^{-xi}),$$

where again, x may be real or complex.

One of the first tasks of complex analysis is to check that all the preceding manipulations involving infinite series are allowable. As was made clear in chapter 2, infinity must be handled with care, no less so when complex numbers are involved.

Integration* turns out to be remarkably different when performed for complex functions. Of course, because complex numbers are two dimensional, you cannot simply integrate from one number a to another number b as with real numbers, as in, for example,

$$\int_0^1 x^2 \, dx = \frac{1}{3}.$$

Instead, you must integrate along a path (in the complex plane). For example, you might want to integrate a function around a circle. What would the answer be if you were to integrate the (complex) function $1/(x - a)$ over a circular path C (where a is a complex constant)? (The integral is written as

$$\int_C \frac{1}{x - a} \, dx.$$

Such integrals are sometimes referred to as *line integrals*.) The result is quite unexpected, especially to those who remember how difficult integration can be for real functions. If the number a corresponds to a point in the complex plane *inside* the circle C, the answer is $2\pi i$; if the num-

---

*If you are not familiar with this concept in the calculus for real numbers, you can simply skip the rest of this section and the other occasional references to integration later in the chapter.

ber a lies *outside* the circle, the answer is 0. The surprising thing about this is that the size and position of the circle are not relevant and that the only way in which the constant a affects the answer is in whether it lies inside the circle or outside it. (Although this particular example was chosen to provide such a striking result, it is typical of how complex functions take on a life of their own, quite different from what we have come to expect from real numbers.)

More surprises are in store when complex-function theory is applied to the study of the natural numbers, as will become clear as this chapter unfolds. (It is worth noting that the integral just mentioned plays a significant role in this study—though not one that can be explained in the account given here.)

A Farey Story

A fraction h/k is called a *proper fraction* if it lies between 0 and 1 and if h and k have no common factors. For example, 1/2, 3/4, and 7/8 all are proper fractions, but 2/4, 3/9, and 3/2 are not. For any number n, the *Farey sequence of order n, F_n,* is the sequence of all proper fractions with denominators that do not exceed n, together with the "fraction" 1/1, arranged in increasing order. For example, F_5 is the sequence

$$\frac{1}{5}, \frac{1}{4}, \frac{1}{3}, \frac{2}{5}, \frac{1}{2}, \frac{3}{5}, \frac{2}{3}, \frac{3}{4}, \frac{4}{5}, \frac{1}{1},$$

and F_7 is the sequence

$$\frac{1}{7}, \frac{1}{6}, \frac{1}{5}, \frac{1}{4}, \frac{2}{7}, \frac{1}{3}, \frac{2}{5}, \frac{3}{7}, \frac{1}{2}, \frac{4}{7}, \frac{3}{5}, \frac{2}{3}, \frac{5}{7}, \frac{3}{4}, \frac{4}{5}, \frac{5}{6}, \frac{6}{7}, \frac{1}{1}.$$

It is not clear who first thought of looking at such sequences. The first person to have proved genuine mathematical results concerning them seems to have been a man called C. Haros, in 1802. J. Farey stated one of Haros's results without proof in an article written in 1816, and when Cauchy subsequently saw the article, he discovered a proof of the result and credited the concept to Farey, thereby giving rise to the name *Farey sequence.*

The result proved by Haros and then stated by Farey is that if you take any three successive fractions from a Farey sequence, a/d, b/e, c/f, then $b/e = (a + c)/(d + f)$. For example, the tenth, eleventh, and twelfth members of F_7 are 4/7, 3/5, and 2/3, and

$$\frac{4 + 2}{7 + 3} = \frac{6}{10} = \frac{3}{5}.$$

The other result that Haros proved is that if a/c, b/d are successive members of a Farey sequence, then $bc - ad = 1$. Taking F_7 as an example again, the sixth and seventh members are 1/3 and 2/5, and

$$2 \times 3 - 1 \times 5 = 6 - 5 = 1.$$

It is possible to deduce either of the preceding two statements from the other, which means that in order to prove them both, you need prove only one of them. Both present quite a challenging exercise in algebraic manipulation. (If this is not to your taste, you can at least check the results for a few other Farey sequences.)

For any number n, let $A(n)$ denote the number of terms in the Farey sequence F_n. Thus $A(5) = 10$ and $A(7) = 18$. Suppose you now take the interval from 0 to 1 on the real line and divide it into $A(n)$ equal segments (see figure 47). The points that divide the interval in this way are the points $1/A(n)$, $2/A(n)$, $3/A(n)$, and so on up to $(A(n) - 1)/A(n)$. Since the terms in the Farey sequence F_n are unequally spaced between 0 and 1, many of the numbers in the sequence will not coincide with the equally spaced points dividing up the interval. Let d_1 be the amount

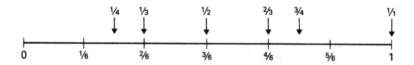

FIGURE 47   The Farey sequence F_4. The members of this sequence are indicated by the arrows, showing their positions relative to the five points that split the interval from 0 to 1 into six equal pieces. The numbers $d_1, d_2, \ldots, d_5$ measure the difference between the Farey fractions and the respective division points. $D(4)$ is the sum of these differences.

by which the first term of the Farey sequence F_n differs from $1/A(n)$, and d_2 the amount by which the second term differs from $2/A(n)$, and so on, up to $d_{A(n)-1}$. (It does not matter which of each pair is greater—the difference between them is what counts.) Then let $D(n)$ be the sum of all the numbers $d_1, d_2, \ldots, d_{A(n)-1}$.

As a simple example, F_4 consists of the numbers

$$\frac{1}{4}, \frac{1}{3}, \frac{1}{2}, \frac{2}{3}, \frac{3}{4}, \frac{1}{1}.$$

Thus $A(4) = 6$. The points that split the interval from 0 to 1 into six equal segments are $1/6$, $2/6$, $3/6$, $4/6$, and $5/6$. So d_1 is the difference between $1/4$ and $1/6$, namely, $1/4 - 1/6 = 1/12$; d_2 is the difference between $1/3$ and $2/6$, which is 0; d_3 is the difference between $1/2$ and $3/6$, also 0; likewise $d_4 = 0$; and $d_5 = 5/6 - 3/4 = 1/12$. So,

$$D(4) = \frac{1}{12} + 0 + 0 + 0 + \frac{1}{12} = \frac{1}{6}.$$

At this stage you should try working out $D(5)$ for yourself.

In a paper published in 1924, J. Franel and Edmund Landau investigated the behavior of the function $D(n)$ as n ranges over all the natural numbers. (They carried out their investigation using algebraic techniques rather than by calculating lots of Farey sequences arithmetically.) In particular, they considered the statement that if r is any real number greater than $\frac{1}{2}$, there is some constant C such that $D(n)$ is always (i.e., for every n) less than Cn^r. They proved that this deceptively simple-looking statement is equivalent to (i.e., is another way of expressing) what is to this day generally acknowledged by professional mathematicians to be the single most important unsolved problem in the field: the *Riemann hypothesis*.

The Most Important Unsolved Problem in Mathematics

Ask any professional mathematician what the single most important open problem in the entire field is, and you are almost certain to receive the answer "the Riemann hypothesis." Obviously, the great British mathematician G. H. Hardy (see chapter 4) thought so. When faced with a ferry crossing from Scandinavia to England on a day when the weather conditions on the North Sea were unusually fierce, he sent a postcard to a colleague (no doubt with the origins of Fermat's last theorem in mind) bearing the message: "Have proved the Riemann hypothesis. Yours, G. H. Hardy." Hardy reasoned that God could not possibly allow him to die with the undeserved credit of proving so important a result and would therefore ensure his safe return home. In the event Hardy (who was, by the way, a devout atheist) got across safely, and "Hardy's last theorem" did not come into being. The Riemann hypothesis remains unproved to this day.

The story begins around 1740, when Euler introduced into mathematics the *zeta function,* defined for real numbers s greater than 1 by the infinite sum

$$\zeta(s) = \frac{1}{1^s} + \frac{1}{2^s} + \frac{1}{3^s} + \frac{1}{4^s} + \ldots .$$

(ζ is the Greek letter zeta; the term *zeta function* has no more significance other than the fact that it is generally denoted by this letter.) For s less than or equal to 1, the infinite sum has an infinite answer, so $\zeta(s)$ is not defined for such an s. But for any s greater than 1, the infinite sum has a definite, finite value. Euler proved that for any such s, the value $\zeta(s)$ is equal to the infinite product

$$\frac{1}{1 - (1/2)^s} \times \frac{1}{1 - (1/3)^s} \times \frac{1}{1 - (1/5)^s} \times \frac{1}{1 - (1/7)^s}$$

$$\times \frac{1}{1 - (1/11)^s} \times \ldots ,$$

where the (infinite) product is over all numbers of the form

$$\frac{1}{1 - (1/p)^s},$$

where p is a prime number. Two things make this result an amazing one. First, it shows that the zeta function is closely related to those decidedly basic and concrete entities, the prime numbers. Second, the relationship between the two involves the infinite to a very great degree. Clearly, there is more to the prime numbers than meets the eye!

For the next episode in the saga, we switch (temporarily) from the zeta function to the distribution of the prime numbers themselves. As you proceed up through the natural numbers, at first the primes appear to be very numerous (e.g., exactly half the first ten numbers beyond 1 are prime), but later they begin to thin out. And yet there seems to be no overall pattern in the way that the primes are distributed among the rest of the numbers. For instance, between 9,999,900 and 10,000,000, there are nine prime numbers, but of the next hundred integers, from 10,000,000 to 10,000,100, there are only two: 10,000,019 and 10,000,079. In fact, there are runs of integers of all lengths that contain no prime numbers. (For any value of N, the run from $N! + 2$ to $N! + N$ cannot include any primes, as is easily seen.) Another example of the seemingly random way in which the primes appear is that there are many examples of pairs of "twin primes," primes that differ by 2, such as 3 and 5, 11 and 13, or 10,006,427 and 10,006,429, and these twin-prime pairs seem to occur in a random fashion. (It is thought that there are infinitely many such twin-prime pairs, but this has never been proved.)

Nevertheless, there is a form of order underlying the seemingly chaotic manner in which the primes appear. It concerns the behavior of the function $\pi(n)$, which tells you the number of primes less than or equal to n (see chapter 1). In his book *Essai sur la théorie des nombres* (1798), Anne-Marie Legendre observed that $\pi(n)$ is approximately equal to the number

$$\frac{n}{\log_e n - 1.08366}.$$

(Throughout this chapter, $\log_e n$ denotes the natural, or base e, logarithm of n.) There is nothing special about the number 1.08366 here. Legendre obtained his result by examining tables of primes up to 400,000 and simply chose this number to give as good an answer as possible.

At about the same time as Legendre was working on his book, the 14-year-old Gauss also began to investigate the function $\pi(n)$. He observed (though he did not publish the fact until 1863) that $\pi(n)$ is approximated by the number $n/\log_e n$ and also by the number

$$\mathrm{Li}(n) = \int_{2}^{n} \frac{1}{\log_{e} x}\, dx.$$

(The function Li is the "logarithmic integral" function. Don't worry if its definition means nothing to you; it is included for "completeness" only.) Table 2 gives the values of the various approximating functions for values of n up to 100,000,000. According to this table, it appears that $\mathrm{Li}(n)$ gives a much better approximation of $\pi(n)$ than do either of the others, and indeed in 1896 it was shown by Charles de la Vallée Poussin that this is so for all values of n from a certain point on.

In fact—and this is a slight but interesting digression—according to table 2, it seems that $\mathrm{Li}(n)$ is always slightly larger than $\pi(n)$, and if you were to tabulate further, it is virtually certain that this would continue to be what you would observe. It would be a skeptical person indeed who did not then conclude that $\mathrm{Li}(n)$ always approximates $\pi(n)$ on the large side. But if you did come to such a conclusion, you would be wrong! In 1914, the English mathematician J. E. Littlewood (a colleague of G. H. Hardy's) showed that the difference $\mathrm{Li}(n) - \pi(n)$ changes from positive to negative *infinitely many times* as n runs up through the positive integers, so there will certainly be values of n for which $\mathrm{Li}(n)$ is smaller than $\pi(n)$. Indeed, such an n will, as S. Skewes showed in 1955, have to appear somewhere before the number

---

TABLE 2   The distribution of primes. This is an expansion of table 1 (in chapter 1) and shows the number of primes $\pi(n)$ smaller than n for various values of n, along with three classical approximating functions to $\pi(n)$.

n	$\pi(n)$	$\dfrac{n}{\log_e n - 1.08366}$	$\dfrac{n}{\log_e n}$	$\mathrm{Li}(n)$
1000	168	172	145	178
10000	1229	1231	1068	1246
100000	9592	9588	8686	9630
1000000	78498	78534	72382	78628
10000000	664579	665138	620420	664918
100000000	5761455	5769341	5428681	5762209

$$10^{10^{10^{10^{10^3}}}}$$

a number of incomprehensibly large magnitude. Considerably smaller, but still well beyond human grasp, is the number 1.65×10^{1165} that R. Sherman Lehman showed in 1966 could be substituted in place of Skewes's number (in the sense that $Li(n) - \pi(n)$ changes sign somewhere below that number). Smaller still is the number 6.69×10^{370}, and in 1986 H. J. J. te Riele demonstrated that a sign change occurs for some n below this bound. But these still are enormously large numbers, which is what has led people to conclude that it may never be possible to discover an actual number n for which $Li(n)$ is less than $\pi(n)$. (Certainly, a computer search made as far as 1 billion (10^9) failed to produce such a number.)

Now we return to the main story. In 1896, the Frenchmen Jacques Hadamard and de la Vallée Poussin independently proved conclusively what all the evidence suggested: that as n increases, the value of $n/\log n$ comes closer and closer to $\pi(n)$ in such a way that no matter how close you want the approximation to be—other than exact equality—you can achieve such accuracy by choosing a large enough n. (It follows that $Li(n)$ also becomes "arbitrarily close" to $\pi(n)$ as n gets larger.) This celebrated result—which shows once and for all that there is a definite, mathematical pattern to the way the prime numbers appear—is known as the *prime number theorem*. (Note, however, that the pattern behind the primes involves infinity and the concepts of the calculus.) The work of both mathematicians was based on a remarkable eight-page paper written by the German Bernhard Riemann in 1859, the title of which translates as "On the Number of Primes Less Than a Given Magnitude." In this, his only paper on number theory, Riemann instigated some lines of research that to this day are still proving extremely fruitful. Indeed, the publication of his paper could be said to mark the beginning of the entire field of analytic number theory, in which the powerful techniques of the calculus are brought to bear on problems about the natural numbers.

The key idea introduced by Riemann in his study of the distribution of the primes was to extend the zeta function $\zeta(s)$ so that instead of being restricted to the real numbers greater than 1, s is allowed to be any complex number $s = a + bi$ (except $s = 1$). This cannot be done simply by allowing s to be complex in the original Euler definition,

$$\zeta(s) = \frac{1}{1^s} + \frac{1}{2^s} + \frac{1}{3^s} + \frac{1}{4^s} + \dots .$$

Instead you must use a rather sophisticated technique known as *analytic continuation* (which I will not describe here). For the benefit of those readers who are able to understand the various notions involved (and no doubt for the bafflement of the rest), the formula that Riemann obtained for his extended zeta function in terms of a line integral is

$$\zeta(s) = \frac{\Pi(-s)}{2\pi i} \int_C \frac{(-x)^s}{e^x - 1} \frac{dx}{x},$$

where the integral is taken over the path C which runs left from ∞ along the positive real axis, stops just short of the origin, circles around the origin in a counterclockwise direction, and then goes back along the positive real axis to ∞. The product $\Pi(s)$ is given by

$$\Pi(s) = \lim_{N \to \infty} \frac{N!}{(s+1)(s+2) \dots (s+N)} (N+1)^s,$$

for all s that are not negative integers.

The extended function is known as the *Riemann zeta function*. It is of fundamental importance throughout number theory, and entire books have been written about it. (One is Harold Edwards's 300-page *Riemann's Zeta Function*. As you might have gathered from the definition of the extended zeta function just given, such books are strictly for the expert.)

For s equal to any one of the numbers -2, -4, -6, and so on, $\zeta(s)$ is equal to 0. Another way of saying this is that the even negative integers are *zeros* of the zeta function. There are infinitely many other complex numbers s for which $\zeta(s) = 0$, all of which have their real part between 0 and 1 (i.e., they are of the form $s = a + bi$, where a is between 0 and 1). The *Riemann hypothesis* is a conjecture Riemann made in his paper about these complex zeros of the zeta function. He suggested (on hardly any real evidence, as far as anyone can see) that all complex zeros of the zeta function have their real part exactly equal to $\frac{1}{2}$ (i.e., if $\zeta(s) = 0$, then $s = \frac{1}{2} + bi$ for some number b).

Why is this hypothesis so important? Well, as Riemann demonstrated in his paper, there is a very close relationship between the zeros of the zeta function and the properties of the function $\pi(n)$. It was this connection that led Hadamard and de la Vallée Poussin to their respective proofs of the prime number theorem. (Their proofs depended on the *connection*, which is known to be true, and not on Riemann's hypothesis about the zeros, whose truth remains unknown to this day.) The connection also lies behind many other known facts about primes, including the work on Skewes's number mentioned earlier. If the Riemann hypothesis does turn out to be true and the zeros of the zeta function really are so well ordered, then the connection with the function $\pi(n)$ will enable even more information about the prime numbers to be deduced than is currently known, and this is what makes it such an important problem for mathematicians. So just what is known about the likelihood of the hypothesis being true?

Before describing the considerable work that has been done on the Riemann hypothesis since it was first put forward, I should tell you one specific approximation to $\pi(n)$ that involves the zeta function explicitly. (This one requires only the original Euler form of the zeta function.) It is (and again, if you are not familiar with the notation, don't worry, just read on):

$$R(n) = 1 + \sum_{k=1}^{\infty} \frac{1}{k\zeta(k+1)} \frac{(\log n)^k}{k!}.$$

(The sum in this equation is an infinite sum—from $k = 1$ to $k = \infty$.)

Table 3 shows what an astonishingly good approximation to $\pi(n)$ the quantity $R(n)$ is.

The Riemann Hypothesis

What, then, is the evidence to support Riemann's conjecture that if $\zeta(s)$ = 0 for a complex number s, then s is necessarily of the form $\frac{1}{2} + bi$? As mentioned earlier, any complex zero will have its real part between 0 and 1, and $\frac{1}{2}$ lies midway between these two bounds, but Riemann surely had greater evidence than that. Whatever the grounds for his assertion, however, no one alive today knows what they were. As will soon become clear, he was not in a position to sit down and calculate many zeros, and even if he had, the example concerning the sign of the expression $\text{Li}(n)$ − $\pi(n)$ shows just how unreliable numerical evidence can be in mathematics. G. H. Hardy was able (and it was not at all easy) to prove that infinitely many zeros have their real part equal to $\frac{1}{2}$, but this does not

TABLE 3   The distribution of primes. This table begins where table 2 ends. For these larger values of n, the function $R(n)$, which is defined using the Riemann zeta function, provides an extremely good approximation to the prime-distribution function $\pi(n)$.

n	$\pi(n)$	$R(n)$
100000000	5761455	5761552
200000000	11078937	11079090
300000000	16252325	16252355
400000000	21336326	21336185
500000000	26355867	26355517
600000000	31324703	31324622
700000000	36252931	36252719
800000000	41146179	41146248
900000000	46009215	46009949
1000000000	50847534	50847455

preclude there being one that does not (or even infinitely many more that do not). Apart from that, there really is little more to go on other than the numerical evidence accumulated as a result of some prodigious calculations over the years. (Although such evidence can never prove the hypothesis, if it turns out to be false, it is always possible—but un-likely—that numerical computation will reveal a number that demon-strates this falsity. For all it would take to disprove the Riemann hy-pothesis would be the discovery of just one zero whose real part was not equal to $\frac{1}{2}$. This provides one justification for taking a computational ap-proach. Another reason is that it provides an excellent testing ground for new numerical algorithms that may turn out to have other uses.)

There are a number of techniques for computing the values of b for which the complex number $\frac{1}{2} + bi$ is a zero of the zeta function. There also are ways of calculating the total *number* of zeros (as opposed to cal-culating the zeros themselves) whose imaginary part lies within any specified range. By combining two such calculations, it is possible to check the Riemann hypothesis for any given finite range. The first person to adopt a computational approach to the Riemann hypothesis was J.-P. Gram. In 1903, using a standard technique called *Euler-Maclaurin summation* to calculate the zeros, Gram found the first fifteen values of b for which $\zeta(\frac{1}{2} + bi) = 0$. The first ten he computed to six decimal places, the first one being $b = 14.134725$ and the tenth $b = 49.773832$. The remaining five he computed to only one decimal place, the eleventh being $b = 52.8$ and the fifteenth $b = 65.0$. Knowing that any complex zero must have its real part between 0 and 1, Gram went on to show that there are exactly ten zeros whose imaginary part lies between 0 and 50. Since his earlier list of zeros of the form $\frac{1}{2} + bi$ has exactly ten entries whose imaginary part lies within the range 0 to 50, it follows that this list comprises all the zeros with an imaginary part in that range. In other words, Gram's computations proved that the Riemann hypothesis is true in the range 0 to 50. (The "range" in this context always refers to the size of the imaginary part of the zero.)

Using similar (although somewhat improved) techniques, R. Backlund in 1918 verified the hypothesis for all zeros in the range 0 to 200, and (using still further improvements to the method) in 1925 J. J. Hutchinson raised the upper limit to 300. In 1936, E. C. Titchmarsh and L. J. Comrie used an improved method devised by Carl Siegel to compute 1041 zeros, all of which were of the form hypothesized by Riemann.

After World War II, electronic computers were brought in to work on the problem. During the 1950s, using the Siegel method for computing zeros of the form $\frac{1}{2} + bi$ together with a new method suggested by Alan Turing for determining the number of zeros in a given range, Derrick Lehmer began to take the search much further, and by 1966, R. S. Lehman had taken the number of computed zeros to 250,000. Within a few years of that, J. B. Rosser and his colleagues took the total to 3,500,000. By 1983, after work by Jon van de Lune and Herman te Riele, the entire region from 0 to 119,590,809,282 had been investigated, and all the (exactly) 300,000,001 zeros in this region had been found to be of the type Riemann predicted. In 1985, van de Lune and te Riele took the calculation as far as the first 1.5 billion zeros, again without finding one that countered the Riemann hypothesis.

Thus all the numerical evidence supports Riemann's hypothesis. If it is false, it must fail for numbers way beyond the range generally considered by anyone other than professional pure mathematicians. Despite all the work and all the evidence, no one really knows whether or not it is true. But as if to caution once again against the temptation to base conclusions on numerical evidence, in the early 1980s a conjecture closely related to the Riemann hypothesis was resolved, and in this case, the numerical evidence turned out to be entirely misleading. The fate of the *Mertens conjecture* should serve as a warning to all who "assume" that the Riemann hypothesis must be true.

The Mertens Conjecture

If you take any natural number n, then, by the fundamental theorem of arithmetic, either n is prime or else it can be expressed as a product of a unique collection of primes. For instance, for the first five nonprimes,

$$4 = 2 \times 2, \quad 6 = 2 \times 3, \quad 8 = 2 \times 2 \times 2,$$
$$9 = 3 \times 3, \quad 10 = 2 \times 5.$$

Of these, 4, 8, and 9 have prime decompositions in which at least one prime occurs more than once, whereas in the decompositions of 6 and 10, each prime occurs once only. Numbers divisible by the square of a prime (such as 4, 8, 9) are called *square divisible*, and numbers not so divisible are called *square free*. (Thus in the prime decomposition of a square-free number, no prime occurs more than once.)

If n is a square-free natural number that is not prime, it is a product of either an even number of primes or an odd number of primes. For example, $6 = 2 \times 3$ is a product of an even number of primes, whereas $42 = 2 \times 3 \times 7$ is a product of an odd number of primes. In 1832, Augustus Ferdinand Möbius introduced the following simple function (denoted by the Greek letter mu and called the *Möbius function*) to indicate what type of prime factorization a number n has. For $n = 1$, let $\mu(n) = 1$, a special case. For all other n, $\mu(n)$ is defined as follows:

If n is square divisible, then $\mu(n) = 0$.

If n is square free and the product of an even number of primes, then $\mu(n) = 1$.

If n is either prime or square free and the product of an odd number of primes, then $\mu(n) = -1$.

For example, $\mu(4) = 0$, $\mu(5) = -1$, $\mu(6) = 1$, $\mu(42) = -1$. You may work out further values for yourself.

For any number n, let $M(n)$ denote the result of adding together all values of $\mu(k)$ for k less than or equal to n. For example,

$$M(1) = \mu(1) = 1,$$

$$M(2) = \mu(1) + \mu(2) = 1 + (-1) = 0,$$

$$M(3) = \mu(1) + \mu(2) + \mu(3)$$
$$= 1 + (-1) + (-1) = -1,$$

$$M(4) = \mu(1) + \mu(2) + \mu(3) + \mu(4)$$
$$= 1 + (-1) + (-1) + 0 = -1,$$

$$M(5) = 1 + (-1) + (-1) + 0 + (-1) = -2,$$

and you can check for yourself that

$$M(6) = -1, \quad M(7) = -2, \quad M(8) = -2,$$
$$M(9) = -2, \quad M(10) = -1, \quad M(11) = -2,$$
$$M(12) = -2, \quad M(13) = -3, \quad M(14) = -2,$$
$$M(15) = -1, \quad M(16) = -1, \quad M(17) = -2,$$
$$M(18) = -2, \quad M(19) = -3, \quad M(20) = -3.$$

(Question: What is the first value of n for which $M(n)$ is zero again? Or positive again?)

All this seems a pleasant enough bit of fun but hardly likely to be related to the most important unsolved problem in mathematics, you might think. (Though having read the section on Farey sequences, you may not be quite so sure.) Well, as will become clear in a moment, the behavior of the function $M(n)$ turns out to be very closely related to the location of the zeros of the Riemann zeta function.

The connection was certainly known to T. J. Stieltjes. In 1885, in a letter to his colleague Charles Hermite, he claimed to have proved that no matter how large n may be, the size of $M(n)$ (i.e., neglecting any minus sign) is always less than $\sqrt{n}$:

$$|M(n)| < \sqrt{n} \qquad (7)$$

(The two vertical bars here are the standard notation for indicating the suppression of any minus sign in the expression within. Thus $|-10| = 10$, $|5| = 5$, and so on.) If what Stieltjes claimed had been true, the truth of the Riemann hypothesis would have followed at once. In fact, the Riemann hypothesis follows from the existence of any constant A such that the inequality

$$|M(n)| < A\sqrt{n}$$

holds for all n. Needless to say in view of what we have seen in the previous section, Stieltjes was wrong in his claim, although at the time this was not apparent. For instance, in 1896, when Hadamard wrote his now-classic and greatly acclaimed paper proving the prime number theorem, he mentioned that he understood Stieltjes had already obtained the same result using the inequality (7), and he excused his own publication on the grounds that Stieltjes's proof had not yet appeared! The fact that Stieltjes never did publish a proof might well suggest that he eventually realized his error. At any rate, in 1897, F. Mertens produced a 50-page table of values of $\mu(n)$ and $M(n)$ for n up to 10 000, on the basis of which he concluded that inequality (7) was indeed "very probable." As a result, this conjecture became known as the *Mertens conjecture*.

In a series of papers appearing between 1897 and 1913, a chap called von Sterneck published additional values of $M(n)$ for selected values of n up to 5 million and found that they too all satisfied the Mertens conjecture. Indeed beyond $n = 200$, they all satisfied the stronger inequality

$$|M(n)| < \tfrac{1}{2}\sqrt{n},$$

but his conjecture that this was always the case was proved to be false in 1960 by Wolfgang Jurkat. The smallest value of n for which $|M(n)| \geq \tfrac{1}{2}\sqrt{n}$ is $n = 7{,}725{,}038{,}629$, which gives the value $M(n) = 43{,}947$.

Then in 1979, Henri Cohen and Andrew Dress computed $M(n)$ for all n up to 7.8 billion and observed that all their values satisfied the inequality

$$|M(n)| < 0.6\sqrt{n},$$

which again seemed to suggest that the Mertens conjecture might be true. (To be fair, I should mention that much other evidence suggested otherwise.) But in fact, it is not true, and in October 1983 Hermann te Riele and Andrew Odlyzko brought eight years of collaborative work to a successful conclusion by proving just that. Their result was obtained by a combination of classical mathematical techniques and high-powered computing.

So how did they set about obtaining their result? Certainly not by finding a number n for which $|M(n)| \geq \sqrt{n}$. To date, no such number has been found, and the available evidence suggests that there is no such n below 10^{30}. Rather, the key to their proof was provided by a result obtained by A. E. Ingham in 1942.

To begin with, note that the inequality $|M(n)| < \sqrt{n}$ can be rewritten as

$$\frac{|M(n)|}{\sqrt{n}} < 1.$$

What Ingham did was to show how to define a certain function $h(x)$ which has the following property: for any (real) number x, no number less than $h(x)$ can be greater than every value of $|M(n)|/\sqrt{n}$. So in order to disprove the Mertens conjecture, you need to find a number x such that $h(x) > 1$. Suppose, for example, you can find an x with $h(x) = 1.06$. Then the Ingham result tells you that no number less than 1.06 is greater than every value of $|M(n)|/\sqrt{n}$. In particular, 1 (which is less than 1.06) is not greater than every value of $|M(n)|/\sqrt{n}$. This contradicts the Mertens conjecture. (But notice that it does not give you a value of n for which the Mertens inequality fails.)

So, in order to dispose of the Mertens conjecture, te Riele and Odlyzko set out to find a value of x for which $h(x)$ is greater than 1. This turned out to be a difficult task.

The definition of the function $h(x)$ involved some very abstract mathematics, including the Riemann zeta function. In particular, the computation of $h(x)$ for a given value of x required the fairly accurate calculation of a considerable number of zeros of the zeta function. (Moreover, Ingham's result for the function $h(x)$ depended on the fact that all the zeros calculated were in accordance with the Riemann hypothesis—though if you expect this hypothesis to be true, you are not likely to worry about this side of things, and in any case, the hypothesis has been checked way beyond the range required to disprove the Mertens conjecture.) These computations were certainly far too time-consuming to be performed by anything but the most powerful computers available. And even when te Riele and Odlyzko had written a program that could perform the calculation of $h(x)$ for any given x, they were faced with the task of finding an x for which the answer was greater than 1. This part was even harder. The function $h(x)$ exhibits a frustrating tendency to produce values that are almost always well below 1. Indeed, by 1979, the best that te Riele had been able to achieve was a value of 0.86. Finding a suitable x was, it seemed, like looking for a needle in a haystack. At the time, te Riele concluded that the problem was outside the range of existing computers.

The breakthrough that led to the solution of the problem came not in the form of new computer technology but from a powerful (and, it transpired, widely applicable) new algorithm discovered by Arien Lenstra, Hendrik Lenstra, and László Lovász in 1981. When applied to the Mertens conjecture, this new technique (which will not be described here) provided just the additional computational facility required to find an appropriate value of x. At that point te Riele and Odlyzko had a method for searching through their haystack.

The first part of the proof was to calculate the zeros of the zeta function that were likely to be required to compute the values of the $h(x)$ function they would encounter in their search. A CDC Cyber 750 computer at the Amsterdam Mathematical Center was run for 40 hours to do this, obtaining 2000 zeros accurate to 100 decimal digits. Then, using the powerful new algorithm just mentioned, a Cray-1 computer at Bell Laboratories was run for 10 hours until a value of x was found for which $h(x)$ was greater than 1. In fact, for the value of x found, $h(x)$ = 1.061545. The Mertens conjecture had finally been laid to rest. For the record, the value of x that did the trick was the following giant, which has 65 digits before the decimal point:

$$-14,045,289,680,592,998,046,790,361,630,399,781,127,$$

$$400,591,999,789,738,039,965,960,762.521505$$

Not only did this result disprove the Mertens conjecture, it also showed that the inequality

$$|M(n)| < 1.06\sqrt{n}$$

is not always valid. But what about the inequality

$$|M(n)| < A\sqrt{n}$$

for other values of the constant A? If for any value of A such an inequality were valid (for all n), the Riemann hypothesis would follow. To prove that an inequality of this form is false using the method adopted by te Riele and Odlyzko, you would have to find a value of x for which $h(x)$ is greater than A. Both te Riele and Odlyzko believe that this is theoretically possible, however large A may be—which means that they believe the function $h(x)$ achieves arbitrarily large values. But, they caution, the best that can be hoped for in practice, given present-day algorithms and computer technology, is to find a value of x for which $h(x)$ is around 1.5. Even reaching 2 is, they conclude, beyond our present capabilities.

So with the Mertens conjecture proved false, where does that leave the Riemann hypothesis? Exactly where it was. Had the Mertens conjecture been true (as Stieltjes had once thought), the truth of the Riemann conjecture would have followed at once. But knowing that the Mertens conjecture is false says nothing about the Riemann hypothesis one way or the other. The two conjectures, Mertens's and Riemann's, are not equivalent.

There is, however, a weakened version of the Mertens conjecture that is equivalent to the Riemann hypothesis, namely, that for any real number r greater than $\frac{1}{2}$, there is a constant A such that the inequality

$$|M(n)| < An^r$$

is valid for all n. In other words, for any power of n greater than the $\frac{1}{2}$ of the Mertens conjecture, you get an inequality of the appropriate type. The falsity of this proposition would, of course, imply the falsity of the Riemann hypothesis (just as its truth would imply the truth of the Riemann hypothesis). But that is an altogether different problem from the Mertens conjecture.

The Bieberbach Conjecture

As it is generally presented to the world, mathematics is a body of cold, impersonal knowledge. Mathematical truth, uniquely, does not vary from minute to minute, from place to place, or from person to person. But mathematics as it is developed is a human endeavor and therefore subject to a whole range of influences. Although at the end of the day the absolute nature of mathematical truth cannot be denied, it can sometimes take some time to arrive at that day's end.

Imagine for a moment that you are one of the world's leading mathematicians. You have been working hard on a particular problem for many years, and on several occasions you have come close to solving it. One day you receive a 385-page typed manuscript that purports to solve your problem. You check the name at the top. Yes, a respectable mathematician. This is not one of the dozens of "crank" papers you routinely receive and immediately consign to the wastebasket. But this person is 52 years of age, and the conventional wisdom has it (on some fairly convincing statistical evidence) that no mathematician produces much of note beyond the age of 40 or so. Moreover, this particular mathematician has in the past claimed to have solved other problems, and in many cases, serious errors have been found in his arguments. And now he is claiming to have solved a problem that in its 70-year history has defeated not only yourself—an acknowledged world expert on this particular problem— but also many other leading mathematicians around the globe (several of whom have at one time or another mistakenly thought for a short while that they had found a solution). A quick glance through the manuscript reveals that your correspondent is using a complicated method that you and nearly everyone else familiar with the problem would regard as most unlikely to be successful.

Faced with such a situation—and no doubt having many other things that you need to get done—what would you do? This is the position in which the American mathematician Carl FitzGerald found himself in the spring of 1984. Had his countryman Louis de Branges really done as he was claiming and managed to solve the Bieberbach conjecture, finding success where so many others had failed? To FitzGerald, it seemed unlikely. ("I was not hoping for the proof to be correct," he wrote later. "Two of my best papers showed the Bieberbach conjecture was at least close to being true. . . . I did not want these results superseded.") Each of the other dozen or so mathematicians across the country to whom de Branges had also sent a copy of his manuscript showed similar skepticism. So no one read it.

But as chance would have it, de Branges had earlier arranged to visit the Soviet Union as part of an exchange program with the United States, and while he was there (from April to June of that year), he was scheduled to lecture to the mathematicians at the University of Leningrad. It was the ideal place to describe his claimed proof. Among the mathematicians in his audience would be I. M. Milin, E. G. Emel'anov, and G. V. Kuz'mina, three acknowledged world experts on the problem. Indeed, what de Branges was in fact claiming was to have proved a conjecture proposed in 1971 by Milin himself and shown by him to imply the Bieberbach conjecture as a consequence. Although they also were skeptical, the Russian mathematicians proved to be a patient audience, sitting through a series of five 4-hour lectures on the purported proof. Their expectation was that an error would be discovered at any moment.

But that moment never came; there was no error. The proof was correct. The next step was to see whether it could be simplified, to reduce it from its daunting length. It could. After some work, the group managed to find some significant modifications to the argument that shortened the proof to a mere thirteen pages. When these thirteen pages were sent out, the rest of the world sat up and took notice—and believed. De Branges had done what most mathematicians dream of but few ever achieve. He had solved a long-standing open problem that everyone had agreed was "hard."

In fact, it is the difficulty of the problem that was the main reason for its fame. As far as is known, the Bieberbach conjecture does not lead to a host of other significant developments, as does (say) the Riemann hypothesis. But it is one of those nice "tidying-up" results that demonstrate the orderly nature of mathematics in general and complex analysis in particular. Here is what de Branges's result says.

Suppose you have an infinite series of the form

$$B = x + a_2x^2 + a_3x^3 + \dots,$$

where x is a complex variable and all the coefficients a_2, a_3, . . . are complex numbers. (There is no mention of an a_1, since the coefficient of x

is 1, that is, $a_1 = 1$.) For a given value of the variable x, it is possible for the infinite series to produce a finite answer (or "sum"). For this to happen, however, the successive terms in the series must grow smaller at a very fast rate—so fast in fact that the effect of there being an infinite number of terms is counteracted.

For example, the series for $\sin x$ given earlier is of the form of B:

$$\sin x = x - \frac{x^3}{3!} + \frac{x^5}{5!} - \frac{x^7}{7!} + \dots$$

(so $a_2 = 0$, $a_3 = -1/3!$, $a_4 = 0$, $a_5 = 1/5!$, $a_6 = 0$, and so on). This infinite series gives a finite answer for any value of x, because the coefficients a_n grow smaller at such a fast rate. (The same is true of the series for e^x, although this is not of the form of B, since it begins with a 1. But if you subtract this initial 1 to obtain the series for $e^x - 1$, you will get a series of the form of B that, like $\sin x$, gives a finite sum for any value of x.)

If, however, the coefficients in the series do not grow smaller so rapidly, the existence of a finite sum may well depend on the size of the number you take as the value of x. Small values of x may lead to a finite answer, larger values to an infinite answer (or no answer at all in some cases). In particular, if x is less than 1, then as n increases the size of x^n will decrease, and consequently there is a chance of obtaining a finite answer. It all depends on the size of the coefficients a_n. Consequently, when mathematicians study series of the form of B, they frequently restrict their attention to the behavior of the series for values of x for which x^n grows smaller as n increases; that is, they consider only those numbers x of absolute value less than 1. (The absolute value of a complex number x is its distance from the origin in the complex plane, denoted by $|x|$. This notation has already been introduced for real numbers x, when it signified the suppression of any negative sign. If you think about it for a moment, you will realize that the two uses are not contradictory. The new use for complex numbers simply extends the old one. By a straightforward application of Pythagoras' theorem for right-angled triangles, if $x = a + bi$, then $|x| = \sqrt{a^2 + b^2}$.)

Thought of in terms of a complex plane, the set of all those complex numbers x for which $|x| < 1$ forms a circular disk of radius 1 with its center at the origin, usually referred to simply as the *unit disk*. (So to say that x lies in the unit disk is just a geometric way of saying that $|x| <$ 1.)

Suppose now that you have a series of the form of B and that this series gives a finite (complex) answer (which I shall denote by $f(x)$) for every value of x in the unit disk. Thus the series determines a function $f(x)$, which to each x in the unit disk assigns a *value* $f(x)$. Now it may be that you can find two different values of x that both give the same answer when you apply the function f. (For instance, suppose that the B series has $a_2 = 1$ and all other coefficients a_n are zero, so that $f(x) = x + x^2$. Then, as you may check for yourself, $f(-1/3)$ and $f(-2/3)$ are both equal to $-2/9$.) If this does not occur (i.e., if different values of x always give different answers), then the function $f(x)$ is said to be *one–one* (because any one value of $f(x)$ comes from only one value of x). Mathematicians give this particular property a name because it is a very useful one—a bit like the requirement of monogamous marriage in most modern societies.

Assuming from now on that our B series (whatever it actually is) does yield a one-one function $f(x)$ that has finite values for every number x in the unit disk, it is possible to picture the action of $f(x)$ geometrically in terms of the complex plane. What $f(x)$ does is associate with every point x in the unit disk another point $f(x)$ somewhere in the complex plane. The set of all the values $f(x)$ obtained in this way will be some region (subset) of the complex plane. It is natural to ask how big this region is. Since the values of x that you start with must come from the unit disk and since the unit disk is small compared with the entire complex plane (which stretches out to infinity in every direction), you might be tempted to think that the region of the values $f(x)$ can also be only a small part of the plane. But remember, both the unit disk and the entire complex plane contain infinitely many points, and as was amply demonstrated in chapter 2, infinite sets do not at all behave in the same way as do the more familiar finite sets. In this case, they most certainly do not. The answer to the preceding question is that the region of values of $f(x)$ may be nearly the entire complex plane! Moreover, you don't need a particularly fancy function $f(x)$ to achieve this. The so-called Koebe function (named after Paul Koebe) does it. This can be calculated either from the formula

$$K(x) = \frac{x}{(1 - x)^2}$$

or else from the B series

$$K(x) = x + 2x^2 + 3x^3 + 4x^4 + 5x^5 + \ldots$$

(Since the Koebe function does come from an infinite series of the form of B and since it is one–one, it is precisely the kind of function we have been looking at.) The set of all values $f(x)$ that the Koebe function gives for x in the unit disk consists of every complex number except for those lying on the real axis to the left of $-\frac{1}{4}$. Geometrically, therefore, this function is pretty well the "biggest" one possible (with "big" referring to the size of the region of values).

Because the Koebe function is a sort of "record holder," you might ask whether this property is reflected in the individual coefficients in its B series in any way. For instance, is each coefficient the largest one possible? That is, suppose you take any series of the form of B that determines a (finite-valued) one-one function $f(x)$ for all numbers x in the unit disk, does it follow that

$$|a_2| \leq 2, \quad |a_3| \leq 3, \quad |a_4| \leq 4, \ldots ?$$

(Or, to put the question the other way around, if your B series has one or more coefficients that *fail* to satisfy the relevant inequality, say if $|a_{163}| = 163.5$, does it follow that either f fails to be one–one or, even worse, there is some number x in the unit disk for which the B series fails to give a finite answer?)

This is the question that the German mathematician Ludwig Bieberbach asked himself early this century, and in a paper published in 1916 he conjectured that the answer was yes. But the only coefficient for which he was able to prove his conjecture was the first one; that is, he did manage to prove that $|a_2| \leq 2$.

And that was where the saga began. The conjecture seemed a reasonable enough one, just the kind of "tidy" result that complex analysis so often yields, and it was sufficiently easy to state to guarantee that there would be no shortage of mathematicians prepared to attempt a proof. But it was to be a long time before that proof was discovered.

After Bieberbach himself, the first result concerning the conjecture was obtained by another German, Charles Löwner, who in 1923 developed a method that enabled him to prove the conjecture for the coefficient a_3 (i.e., he showed that $|a_3| \leq 3$). It was not until 1955 that Paul Garabedian and Marcelo Schiffer, working in the United States, succeeded in proving the conjecture for the next coefficient, a_4. Then in 1968, Roger Pederson and Mitsuru Ozawa skipped over a_5 to prove the conjecture for a_6, and in 1972, Pederson and Schiffer teamed up to dispose of the bypassed a_5. In the same year, Ozawa and Yoshihisa Kubota proved that $|a_8| \leq 8$, missing out a_7, and that was to mark the end of this step-by-step approach.

Slow progress indeed. And moreover, there was no possibility that the conjecture could be proved in its entirety by this process of examining each coefficient one by one (since the conjecture was for every one of the infinitely many coefficients). But there was always a chance that such an approach might lead to just the insight required to solve the complex problem. In the event, this turned out to be true: when Louis de Branges finally proved the Bieberbach conjecture in 1984, his method was indeed based on some of the early work. Which early work? That of Löwner in 1923, the first person to work on the problem after Bieberbach himself. But we are getting ahead of the main story, for while slow progress was being made in attacking the coefficients one by one, another approach was also under way.

The idea was to investigate inequalities of the form

$$|a_n| \leq Cn,$$

where C is some constant, and to see how small C could be made while the inequality was provably valid for all values of n. To prove the Bieberbach conjecture, it would be necessary to show that $C = 1$ was possible, but in the meantime, how close could you get to this value?

The first result of this kind was found by J. E. Littlewood, who in 1925 proved that $C = e$ was possible. (Recall that e is approximately equal to 2.718.) After that, various workers steadily lowered this value. In 1956, Milin (mentioned earlier) showed that you could have $C = 1.243$. In 1972, FitzGerald (the recipient of the 385-page manuscript mentioned at the beginning of this section) lowered this to $C = 1.081$, and in 1978, his student David Horowitz took it down to 1.066. They were getting tantalizingly close, but not close enough. And when the complete solution to the Bieberbach conjecture finally came in 1984, it was (as mentioned earlier) based, not on this "reduce C" approach, but on Löwner's original 1923 work.

To prove the conjecture for a_3, Löwner introduced a partial differential equation whose solutions approximate any function that is one–one on the unit disk. The trick now was to translate the mathematics into physical terms, such as the flow of water along a river. The differential equation represents an expanding flow, and it is possible to "push information" along the flow—in particular estimates of the size of the coefficients in the B series for the function. For his proof, de Branges introduced some auxiliary functions t_1, t_2, . . . to hold the desired information, each t_k being defined by a differential equation involving all previous functions up to t_{k-1}. The proof of the Bieberbach conjecture (in fact of the stronger Milin conjecture alluded to earlier) then reduced to a proof that these t functions satisfy certain conditions. It was, all in all, a decidedly "old-fashioned" approach to the problem, in many ways resembling a huge juggling trick as de Branges struggled to keep all the balls in the air. But it worked.

Suggested Further Reading

An introductory account of complex numbers can be found in many elementary books on mathematics, for example *Sets, Functions and Logic,* by Keith Devlin (Chapman & Hall, 1992). Likewise, there are many sources of more advanced information about complex analysis, such as *A First Course on Complex Functions,* by G. J. O. Jameson (Chapman & Hall, 1970).

Introduction to Analytic Number Theory, by K. Chandrasekharan (Springer-Verlag, 1968), does exactly what its title says—although the subject is, by its very nature, not accessible to the nonmathematician.

Another good, solid introduction to the subject is *Introduction to Analytic Number Theory,* by Tom Apostol (Springer-Verlag, 1976).

The classic book on Riemann's zeta function is *Riemann's Zeta Function,* by H. M. Edwards (Academic Press, 1974).

The solution to the Mertens conjecture is described in the research article "Disproof of the Mertens Conjecture," by A. M. Odlyzko and H. J. J. te Riele, in the *Journal für die reine und angewandte Mathematik* 357 (1985):138–160.

For a brief account of the Bieberbach conjecture and its solution, see the article "The Last 100 Days of the Bieberbach Conjecture," by O. M. Fomenko and G. V. Kuz'mina, in the journal *The Mathematical Intelligencer* 8 (1986):40–47. Another account can be found in Carl FitzGerald's article "The Bieberbach Conjecture: Retrospective," in the *Notices of the American Mathematical Society* 32 (1985):2–6.

9

Knots, Topology, and the Universe

Boy Scouts, Physicists, and Another Book

How can you tell a reef knot from a granny knot? Most Boy Scouts would have no difficulty in answering, but can a mathematician make the distinction? In 1984, some significant developments affected this question.

Do physicists use the right kind of mathematics to study the four-dimensional space–time universe we live in? Until 1982, every mathematician would have said, Yes, of course, it is the *only* kind. But now we know better. There are other, quite different kinds of mathematics that apply to a four-dimensional universe—but only to a four-dimensional universe; no such special treatment is required for two dimensions, or for three, or for five, six, and beyond. Four-dimensional space is special not only because our universe appears to be four dimensional, it is mathematically special as well, but in a totally unexpected way.

These are just two of the developments that have taken place in recent years in the vast area of mathematics known as *topology*, a subject now so big that it would be possible to write an entire book entitled *Topology: The New Golden Age* instead of only this one chapter. Topology, or at least certain aspects of it, now pervades most areas of present-day mathematics, to say nothing of its many connections with modern physics. The subject has already made one explicit appearance in this book, in the discussion of the four-color theorem in chapter 7. (Connections with topics in other chapters were not brought out but are there nonetheless.) And yet topology is barely a century old. Although some of the ideas go back to Euler and Gauss, it was only with the work of Henri Poincaré and others in the latter years of the nineteenth century that topology really got under way.

This chapter deals with just two aspects of topology: *knot theory*, a fascinating though highly specialized part of the subject, and *manifold theory*, the study of the properties of geometric surfaces and generalizations of them. Most topologists would regard manifold theory as the central thrust of their discipline. (One theme that I ignore, although it satisfies all the criteria for inclusion in this book, is catastrophe theory, a form of applied manifold theory developed in the mid-1960s by the topologists René Thom and Christopher Zeeman. My excuse is that a good layperson's account of this subject is easily available elsewhere.*)

Although topology is an extremely difficult subject to pursue properly, a knack for visualizing geometric objects is all that is required for the interested outsider (yourself, presumably) to grasp the general principles. For this reason, the following account takes a highly geometric approach.

What Is Topology?

Topology is a part of mathematics that is at the same time easy to describe and yet very difficult to appreciate properly. The simple description is that topology is the study of those properties of geometric objects that remain unchanged under continuous transformations. At an intuitive level, a *continuous transformation* (known also as a *topological transformation*) can be thought of as subjecting the object to bending, stretching, compressing, or twisting—or any combination of these, it being assumed that the object being deformed (or *transformed*) is ideally elastic and capable of any degree of such manipulation. The proviso is that points in the object that are close together before the transformation is applied remain close together in the transformed object. When properly formulated—and this is by no means easy—this requirement does not, as it might at first seem, conflict with the object

---

* *Catastrophe Theory*, by Alexander Woodcock and Monte Davis (Penguin, 1980).

being stretched to any degree whatever. Rather, it prevents any cutting, tearing, or "gluing" of the object. At least in general terms, it does; what *is* allowed is the cutting of the object to allow for some manipulation to be performed that would not otherwise be possible, followed by the gluing together of the cut edges so that points that started out close together are once again close together. For instance, in order to change, the object illustrated in figure 48(a) into that shown in (b), by means of continuous transformation, it is permissible to cut one of the two interlocked rings and separate it from the other ring, as in (c), and then glue the two cut edges back together again. Because the two objects are thus transformable one into the other, they are said to be *topologically equivalent*. (In fact, for this particular example, cutting is not necessary. It is possible to transform figure 48(a) into (b) by means of stretching and bending alone. Can you see how to do it? The solution is shown in figure 62, at the end of the chapter. This example shows that even "simple stretching" is quite a tricky concept to grasp.)

One of the reasons that the "stretching-cutting-gluing" description of topology is inadequate is that it applies only to the study of surfaces (i.e., two-dimensional objects), whereas topology deals with objects with any number of dimensions—three, four, five, and beyond. But even for surfaces, there are difficulties. Accustomed as we are to thinking in terms of two-dimensional objects situated in three-dimensional space, it is easy to confuse the properties of an object with the properties of the space surrounding it. Indeed, the concept of dimensionality itself often causes confusion. For example, a sphere is regarded in this chapter as a spherical, *two*-dimensional surface, not a solid ball. A point on this surface can move in only two independent directions *on the surface*. However, a sphere can be constructed only in a space having at least three dimensions.

For another example, the sphere and the torus* shown in figure 49 would seem to be topologically distinct (and indeed they are). By no amount of stretching, bending, or cutting and gluing would it seem possible to transform one into the other. (Remember that the only allowable cutting-and-gluing procedure requires gluing together all the points separated by the cut, so you cannot transform a torus into a sphere by cut-

---

*Like a sphere, a torus is to be thought of as a hollow surface.

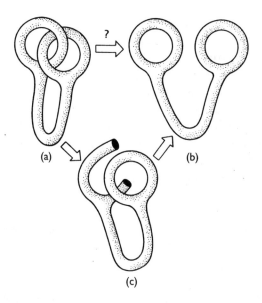

FIGURE 48   A ring puzzle. Imagine the object in (a) made from a perfectly elastic material. Can you transform it so as to unlink the two rings, as in (b)? The most obvious way is to make a cut in one of the rings, as in (c), separate the two rings, and then rejoin the two cut ends. If you join the two free ends in exactly the same way as before they were cut (i.e., without twisting one of the two ends round at all), this procedure will be a legitimate topological transformation of (a) into (b). But it is also possible to transform (a) into (b) without any cutting, simply by manipulating the object in the appropriate manner. Can you see how to do it? A solution is shown in figure 62.

ting across it to form a cylinder and then gluing each end together.) The fundamental difference between the two surfaces would seem to be that the torus has a "hole" in it, wouldn't it? Well, no, not quite. A torus is a smooth surface with no holes at all. If you were constrained to live on a large torus, you could wander all over its surface without ever encountering a hole. The hole is connected with the way this particular surface is situated in three-dimensional space. Or to put it another way, the hole is something in the surrounding space, not in the surface. This does not mean that the hole is irrelevant to the topology of the torus, only that it is not solely a property of the surface itself. There is a topological prop-

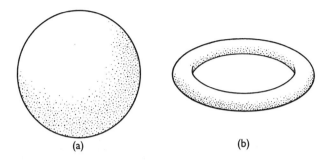

(a)   (b)

FIGURE 49   The sphere (a) and the torus (b). In topology, both of these are thought of as (hollow) surfaces, not solid objects. Most people would agree that it is not possible to transform either of these figures into the other by topological means—that the two surfaces are therefore topologically not equivalent. To prove this, you must show that there is a topological feature of one surface not shared by the other. The "hole" in the torus would seem to be a good candidate. The problem is that the hole is a property not so much of the torus surface but, rather, of the surrounding space. The topological feature that does distinguish between these two figures is the *Euler characteristic*. Although a genuine property of the surface itself, the Euler characteristic of the torus faithfully corresponds to the hole in (or rather "not in") the torus.

erty that the torus possesses that is closely related to the existence of the hole in the surrounding space, and in a moment we shall see what it is. (It serves to distinguish between a torus and a sphere.)

The distinction between a property of a surface and a property of its surrounding space is a subtle one but may become a little clearer when contrasted with the topological notion of an *edge*. Neither a sphere nor a torus has an edge: they both are *closed* (i.e., edge-free) surfaces. A disk has one edge. A disk with a hole in it (such as a compact disk used to store music or data) has two edges. If you glue together the ends of a strip of paper to form a cylindrical band (see figure 50(a)), you will get a surface with two edges. If you give the strip a single half-twist before gluing the ends together (figure 50(b)), you will get a surface with just one edge (make one and see), known as a *Möbius band,* named after August Ferdinand Möbius, whom we met in chapter 8.

If you were to glue together two Möbius bands edge to edge, the re-

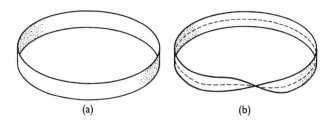

(a)   (b)

FIGURE 50   A cylindrical band (a) and a Möbius band (b). The cylindrical band is an example of a two-edged, orientable surface; the Möbius band is a single-edged, nonorientable surface. To construct a Möbius band from a strip of paper, simply give the paper a single half-twist before gluing the two ends together. The Möbius band has some curious properties. As an example, you could investigate what happens when a Möbius band is cut "in half" along the dashed line shown here. The result is not at all what you might expect. You also could make a similar cut on another Möbius band, but this time one-third of the way in from the edge.

sulting figure would be a Klein bottle (see figure 44 on page 184) However, because you are working in three-dimensional space, this construction is not possible and can be depicted only by allowing the surface to pass through itself. Incidentally, statements of the form "a Möbius band (or Klein bottle) is a surface having just one side" (as in chapter 7) also are a little misleading when used to convey the fundamental ideas of the topology of surfaces, since "sidedness" too depends on regarding the object as sitting inside a space of three or more dimensions (from which lofty standpoint it is possible to look down on the sides of the surface). Discovering that it is impossible to paint what appears to be the two sides of a Möbius band or a Klein bottle using two different colors (try it—at least for a Möbius band—and see what happens) certainly will help you appreciate the unusual nature of these surfaces, but it will not bring out the genuine topological property of the surface, which in this case is not sidedness but "orientability." So from now on, try not to think of surfaces as having "sides."

A surface is said to be *orientable* if the notions of "clockwise" and "counterclockwise" can be distinguished in the following sense. Suppose you draw a small circle on the surface (better still, *in* the surface—no

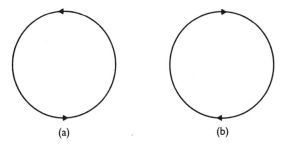

(a)   (b)

FIGURE 51   Orientability. A surface is said to be orientable if it is impossible to turn a counterclockwise-arrowed circle (a) into a clockwise-arrowed circle (b) by transporting the circle over the surface. Orientability is the topological property of surfaces that corresponds to the intuitive notion of "two-sidedness" (as in a cylindrical band); nonorientability corresponds to "one-sidedness" (as in a Möbius band).

"sides," remember) and give it a direction, as in figure 51. Then no matter how you move the circle over (better, *within*) the surface, you will never be able to reverse the direction of the circle. ("Small" as applied to the circle here means small enough to be freely moved all over the surface without getting caught up around any holes, protrusions, or other features.) The cylindrical band (figure 50(a)) is orientable; the Möbius band (figure 50(b)) is *nonorientable*. To see this for yourself, construct your bands from some kind of transparent material (the acetate sheets sold for use with overhead projectors are excellent for this purpose), so that when you draw your directed circles, they can be seen from both "sides" and may therefore be regarded as "in" the surface. Now, starting with a small arrowed circle drawn somewhere on the band, "move" it around the band by making successive copies (including the arrow) at regular intervals. With the Möbius band (see figure 52), when you arrive back at your starting place (you might mistakenly think you are on the opposite "side" from where you began), you will discover that your circle has the arrow pointing the other way around. Your sequence of circles shows how "clockwise" and "counterclockwise" can be interchanged without ever leaving the surface; this is nonorientability. With the cylindrical band, no such reversal is possible; this is orientability.

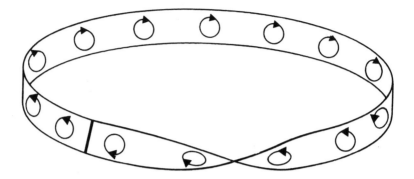

FIGURE 52   Nonorientability of a Möbius band. Construct a Möbius band from a strip of transparent material and follow what happens when a directed circle is moved around the band.

How Do You Do Topology?

A facetious, though highly pertinent, reply to this question would be, With great care. The very general kinds of transformation allowed in topology mean that most of the familiar properties of geometric figures no longer apply. In classical geometry (which may be described as the study of properties of objects that remain unchanged under *rigid transformations*—translation, rotation, and reflection), such concepts as straightness, circularity, angle, length, area, and perpendicularity are used. None of these makes any sense in topology, however, because all of them can be destroyed or altered by means of continuous transformations.

In classical geometry, to verify that two objects are the same, you see if it is (theoretically) possible to move one of them by means of a rigid transformation so as to occupy exactly the same position as the other. This is called *transforming* one object into the other. If such a transformation is possible, the two objects are said to be "the same" or "identical" or, more formally, *geometrically equivalent.* To show that two objects are not the same, you look for a specific geometric property that is different in each one, such as one being bigger than the other or a certain angle being different. All this is so familiar as to appear too trivial to mention. After all, we compare objects in this way all the time. But exactly the same procedure is used in topology, except that the transformations and the distinguishing concepts used are different (and not as familiar to us).

To show that two objects are *topologically equivalent*,* you look for a topological (i.e., continuous) transformation that will turn one into the other. For instance, a triangle and a circle are topologically equivalent, because there is a clear way of deforming one into the other. To demonstrate that two objects are topologically distinct (i.e., nonequivalent), you look for a topological property that only one of them possesses, for example, having an *edge*. A disk has an edge but a sphere does not, and so the two are topologically different. (You cannot continuously deform a disk into a sphere or vice versa.) To distinguish (topologically) between two closed surfaces (i.e., surfaces without an edge), such as a sphere and a torus, another topological property must be found. The "hole" of the torus will not do, since—as we have seen—this in itself is not a topological property of the surface alone. Nor will orientability do, since both surfaces are orientable. (Since the Klein bottle is nonorientable, this concept can distinguish both a sphere and a torus from a Klein bottle.)

One of the first tasks of topology, therefore, is to find enough topological properties to enable us to distinguish between any two nonequivalent objects by showing that one of them has a particular property and the other does not. Any such property must be the same for all objects that are topologically equivalent to the one considered. That is, a topological property must not be changed by a continuous transformation. (If it were, it would not be a topological property.) For this reason, such properties are often called *topological invariants,* a name that is somewhat better than topological *property*, since many invariants are numerical ones. Orientability is one topological invariant, as is the number of edges of a surface. (A surface with one edge cannot be equivalent to one with three edges.) There is one further topological invariant that, together with the two just mentioned, suffices to distinguish among all nonequivalent surfaces. But before we look at it, I should say something about the use of phrases such as "*the* sphere" or "*the* Klein

---

*Mathematicians use the words *same* and *identical*, but for beginners, they carry too strong a geometrical connotation and so I avoid using them here wherever possible.

bottle," which mathematicians use when they do topology. Because any two spheres (say) are topologically equivalent (they can differ only in position and size, neither of which is a topologically invariant), from the topologist's point of view they are "identical," and hence it makes sense to speak of "*the* sphere," even if the object you are actually looking at resembles a long sausage. If two spheres are being considered, it would be incorrect to use a definite article, but when your intention is simply to convey the *idea* of a sphere, the phrase "the sphere" makes sense.

And so to that third invariant of surfaces. In fact, you have met it already, it was introduced and used in chapter 7. As shown in chapter 7, for any surface, the number $V - E + F$ obtained from the number of vertices (V), edges (E), and faces (F) in a map or network covering the (entire) surface is independent of the actual network chosen and its location on the surface. Moreover, it can be proved (and it is certainly easy to believe) that this quantity remains unaltered by any continuous transformation of the surface and so is a topological invariant of the surface. It is called the *Euler characteristic* of the surface.

For the sphere, the Euler characteristic is 2 (i.e., $V - E + F = 2$ for any network covering the surface of any object topologically equivalent to a sphere). For the torus, it is 0, and hence the torus and the sphere are topologically distinct. Incidentally, it is the Euler characteristic that is related to the "hole" of the torus. The double torus, for which there are two holes, has Euler characteristic -2; the treble torus has Euler characteristic -4; and so on; a general formula is given later. For the Klein bottle, the Euler characteristic is, as for the torus, 0, so this invariant does not distinguish between these two objects (but as has already been mentioned, orientability does the job in this case). The three invariants—number of edges, orientability, and Euler characteristic—are enough to distinguish all (two-dimensional) surfaces.

Knot Topology

Some topologists might say that knot topology is not "topology," and in a sense they are right. Knot theory is a kind of topology, but a very special kind. In knot theory, the objects of interest (*knots*) are, of necessity, objects in three-dimensional space. In two dimensions, it would be impossible to construct a knot—there is not enough "room" to wrap the string, or whatever you are using, around itself. In four or more dimensions, there is too much "room"—any knot would immediately "fall out" to leave an unknotted string. Moreover, the kind of topological deformations that may be performed on knots specifically exclude the cutting and gluing permitted in a general topological transformation. (The reason for this restriction should be evident.)

So what is knot theory? Precisely what its name suggests: a mathematical study of knots. Two simple examples of knots are shown in figure 53(a) and (b): the overhand knot and the figure-of-eight knot. If you form either of these two knots from a length of string, you will discover that they really do knot the string; they do not simply "tangle" it. The distinction is that a knot can be undone only by a process that involves pulling a free end through a loop at some stage, whereas a tangle can be removed even if the two ends of the string are held fixed. Moreover, the

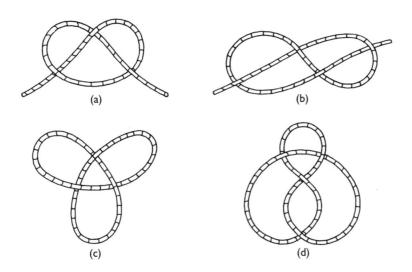

FIGURE 53   Basic knots: (a) overhand, (b) figure-of-eight, (c) trefoil, and (d) four-knot. The first two are types of knot that you might construct using a length of string. For a mathematical study, however, the free ends must be joined together to form a closed loop. Joining the ends of (a) produces (c), and joining the ends of (b) produces (d). Drawings (c) and (d) are examples of what are known as knot diagrams.

overhand knot and the figure-of-eight knot seem to be quite different: one cannot be transformed into the other except by threading or un-threading a free end. (If both ends are fixed, no amount of manipulation will turn an overhand knot into a figure-of-eight knot.)

Despite the apparent simplicity of things so far, however, we are already treading on thin ice. True enough, playing with a piece of string for half an hour or so might fail to unknot the string or to convert one knot to another. But that does not prove that the string is knotted or that two knots are really different. You may simply have failed to find the correct sequence of moves. (Remember the puzzle in figure 48 on page 226?) And what about very complicated-looking configurations? When a long piece of string or wire gets into a hopeless-looking tangle (as happens with frustrating regularity to the cables on electric lawn mowers), it is impossible to tell whether it is merely tangled (in which case, it can be straightened out by carefully pulling the ends) or genuinely knotted. The complicated-looking jumble of wire might well conceal no knot at all or nothing more than a simple overhand knot. And of course, magicians make extensive use of impressive-looking "knots" that do not actually "knot" the rope at all. A potentially very confusing state of affairs indeed and one that requires a decidedly careful approach—a systematic, mathematical approach. This is what knot theory sets out to do.

Since the object of knot theory is to investigate "knottedness," the first thing to do is to get rid of those free ends. Although free ends are essential when you actually want to *tie* a knot, their presence would prevent a mathematical study from ever getting off the ground, since both tying and untying are (cutting-free) topological processes. In other words, from a topological point of view, no string with free ends can be knotted. The simplest solution is to join the free ends together to create a closed (and presumably knotted) loop, which is what is done in knot theory. When the free ends of the two knots in figure 53(a) and (b) are joined, the resulting (mathematical) knots are the *trefoil* and *four-knot*, respectively, shown in figure 53(c) and (d).

A *knot*, then, is just a closed loop (of string, rope, or whatever). (As a special case, this definition includes the simple, "unknotted" closed loop. Known as the *trivial knot*, it is somewhat analogous to the number 0 in arithmetic or the empty set in set theory.) Two knots are said to be *equivalent* if one can be transformed into the other by a topological process that does not involve cutting. The principal aim of knot theory is to find a collection of *knot invariants* that is sufficient to distinguish between any two nonequivalent knots (in much the same way as orientability and Euler characteristic serve to distinguish between any two nonequivalent closed surfaces). In particular, it should be possible to determine whether a given knot really is knotted or whether it is equivalent to the trivial knot (i.e., a simple loop of string).

The first such investigations seem to have been made by the ubiquitous Gauss. Certainly a student of Gauss by the name of Listing devoted a large part of his monograph *Vorstudien zur Topologie* (1847) to the topic, after which a lot of work was done on the subject. (It was not until around 1910 that Max Dehn managed to prove that there is such a thing as a nontrivial knot. Until then, it was still a theoretical possibility that no genuine knots existed to be studied.)

Much of the early work, carried out by T. P. Kirkman P. G. Tait, C. N. Little, and others between 1870 and 1900, concerned the classification of the so-called prime knots (which, as you might imagine, are knots that cannot be split into two simpler knots). These were classified according to their *crossing number*. To obtain the crossing number of a knot, you first lay it out flat on a level surface, either physically if your knot is made of string or else by means of a "projection" if it is a conceptual knot. You then manipulate it so that it has as few crossings as possible, and at no single point do three or more lengths of the knot cross over one another. (A crossing is just a point where the string crosses over itself.) Then the total number of points where the string crosses over itself is the crossing number of the knot, which is an invariant of the knot. For example, both of the knots shown in figure 53(c) and (d) are "laid out" according to this procedure, and so by inspection you can see that they have crossing numbers of 3 and 4, respectively. (These knots are also prime.)

Because it is a knot invariant, the crossing number may be used to distinguish between nonequivalent knots: if two knots have different crossing numbers, they cannot be equivalent. The concept is, however, not as useful as might be imagined. For one thing, many different knots can have the same crossing number, so the invariant often fails to detect nonequivalence. In addition, it can be quite hard to calculate the crossing number, since you must display the knot so that it has no superfluous loops or crossings, and even for simple knots, it usually is not apparent when this has been achieved.

Nevertheless, by the end of the nineteenth century, many prime knots with crossing numbers up to 10 had been distinguished and tabulated. What was not known was whether the tables included all the knots of each of the crossing numbers or whether there were any redundancies in the table in the sense that apparently different but in fact equivalent knots were included as separate entries. In 1927, J. W. (James) Alexander* and G. B. Briggs examined this question of redundancy and managed to prove that there were no duplications in the tables up to and including crossing number 8. Their methods were almost as successful for knots with crossing number 9 as well—only three pairs could not be distinguished by means of the properties that they were using. Subsequent work by K. Reidemeister took care of these outstanding three pairs. Distinguishing all the tabulated knots with crossing number 10 was accomplished only as recently as 1974, by K. A. Perko. As for the question of finding new knots not included in the old nineteenth-century tables, hardly any progress was made until 1960 when John Horton Conway invented a new and more efficient notation for knots. This enabled him not only to discover some prime knots that the earlier workers had missed but also to extend the tables to cover prime knots with crossing number 11. There are 801 prime knots with crossing number at most 11, consisting of one each with crossing numbers 3 and 4, two with crossing number 5, three with 6 crossings, seven with 7, twenty-one with 8, forty-nine with 9, one hundred sixty-five with 10, and five hundred fifty-two with crossing number 11.[†] By 1998, knot theorists had tabulated all prime knots with 16 or fewer crossings. There are exactly 1,701,936 of them altogether. The honor for taking the tabulation to this dizzy height goes to Jim Hoste, Jeff Weeks, and

---

*This is not the same Alexander who is said to have "undone" the famous Gordian knot. That was Alexander the Great, in 333 B.C., and he used the method of slicing through it with a sword, which is forbidden to present-day knot theorists.

[†]The prime knots with crossing numbers 3 and 4 are the two knots shown in figure 53(c) and (d). For pictures of prime knots with up to 9 crossings, see the paper "On Types of Knotted Curves," by J. W. Alexander and G. B. Briggs, in the journal *Annals of Mathematics* 28 (1927):562–586. For knots with 10 crossings, see "On the Classification of Knots," by K. A. Perko, in the journal *Proceedings of the American Mathematical Society* 45 (1974):262–266. For an account of the tabulation of all prime knots with 16 or fewer crossings, see the article "The First 1,701,936 Knots," by Jim Hoste, Morwen Thistlethwaite, and Jeff Weeks, in the magazine *Mathematical Intelligencer* 20, no. 4, Fall 1998, pp. 33–48.

Morwen Thistlethwaite, the first two working together and Thistlethwaite independently. (It is largely because the independent work of two teams led to the same tables that mathematicians are confident that the tabulation is complete and correct. Although both teams used computers, their methods were different.)

By far the most significant advance in the mathematical theory of knots was the realization that with every knot can be associated a certain group—the so-called knot group (for that knot). (See chapter 5 for an introduction to the mathematical concept of a group.) This is one of those marvelous occasions in mathematics when the concepts and results from one part of the subject are discovered to be useful in another, in this case the application of group theory to knot theory.

The idea behind the construction of the knot group is a simple one and can be explained by using the trefoil as an example (see figure 54). Start by choosing some point X not on the knot. (Exactly where X is is not important. The final group you get will be the same wherever it lies.)

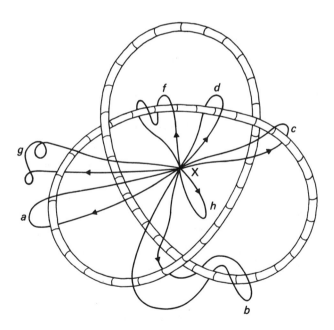

FIGURE 54 Construction of the knot group for the trefoil knot (see the text for details).

To define a group, you must do three things: (1) define the objects that make up the group, (2) establish how two objects in the group are combined together to produce a third object in the group, and (3) check that the various group axioms (see p. 123) are satisfied.

The objects forming the group (for the trefoil as our example) are directed (i.e., arrowed) paths that start at the point X, wind around the knot to various degrees, and finish at the point X. (The paths may not pass through the material of the knot.) But not all such paths are included in the group. Paths having superfluous loops (i.e., loops that can be untwisted or cut out without affecting the way the path winds around the knot) are omitted. Thus path g in figure 54 is not in the group, whereas path a is. Also, when two paths can be deformed into each other (without "passing through" the material of the knot), such as paths a and b, only one of them is included in the group. But note that paths c and d are not such a pair because they pass around the knot in opposite directions, and direction is an important factor. One special path that is also included in the group is the trivial "path," of zero length. Any path that does not wind around the knot at all, such as path h, can be deformed into this trivial path and hence is not included in the group. And that is it. Note that this definition produces a group containing infinitely many objects, since there will be paths that wind around a given length of the trefoil any finite number of times before returning to X, and they will be quite distinct (i.e., not deformable into one another).

Having established what the objects of the group are (and the fact that these may not be the kinds of objects that you usually think of as forming a group shows just how wide-ranging the concept of a group is), the next step is to decide how two such objects are combined, or "multiplied together." Given two (directed) paths p and q, take the path that consists of p followed by q. It will always be possible to deform this combined path to a path that has been put into the group, because all paths are put into the group except those that can be deformed into one already there. It is this path that will be the "product" $p * q$ of the two objects p and q in the group. For example, in figure 54, $d * d = f$, as is easily seen. In combining two paths, opposite directions cancel each other out to leave no path at all. For example, paths c and d cancel out, a fact that may be written as $c * d = e$, or $d * c = e$, if we use e to denote the trivial path.

Finally, we have to check that these definitions produce a system that satisfies the group axioms. I'll leave you to do this part. The identity element of the group is, fairly obviously, the trivial path *e*. Incidentally, for the trefoil at least, the knot group is not commutative, and you should be able to convince yourself of this fact, too.

The knot group is (and this at least seems clear in view of the definition, allowing as it did for deformations of all the paths) an invariant of the knot. Two equivalent knots will give rise to the "same" group; that is, their knot groups will be exact copies of each other in regard to their group behavior, although the precise paths that are put into the two groups will very likely be different. (For finite groups, one group will be an "exact copy" of another if their tables—see chapter 5—are the same. For infinite groups, as here, the mathematical notion of an *isomorphism* is required in order to make this precise.) Hence the knot group may, like the crossing number, be used to distinguish between nonequivalent knots. The knot group is, however, a much more powerful invariant. Indeed, the knot group manages to capture so much of the knot's "knotty structure"—as seems clear from its definition—that it is only rarely that two different knots turn out to have the same knot group. (But the familiar reef and granny knots are such a pair, so there are no grounds for complacency.) Hence the knot group promises to provide an excellent means of telling knots apart. But there is a problem: how do you obtain a sufficiently simple algebraic description of the group? (Remember, the knot group is an infinite, abstract mathematical structure.) Using the description just outlined to define the knot group is obviously not the way to do it because that is even more complicated than the knot itself! Fortunately, this problem turns out to have a solution: in 1910, Dehn discovered a way of obtaining from a diagram of any knot a simple and concise algebraic description of the associated knot group.

This means that the knot group does indeed provide an excellent, practical means of classifying knots. It allows the very powerful techniques of group theory to be brought to bear on the problem, sometimes leading to the development of other useful invariants (one of which, the Alexander polynomial, is described later). In the meantime, let me offer one simple direct application of the knot group. Consider the following question: is it possible to untie a given knot (i.e., make it equivalent to a circle) by introducing an additional knot in the string and then unraveling the result? The answer is no. Tying an additional knot in the string creates a bigger and more complicated knot group than the original one, whereas the knot group of the trivial knot (i.e., the circle) is the same as the group formed by the integers with the operation of addition. Since the knot groups are different, the knots cannot be equivalent.

Another way of looking at knots was discovered in 1935 by Ladislav Seifert. He devised a way to construct, for any given knot, an orientable (i.e., "two-sided") surface that has the knot as its only edge. Now, a standard result in the topology of surfaces (see the next section) is that any one-edged, orientable surface is topologically equivalent to a disk with a certain number of "handles." A "handle" is added to a disk by cutting two holes in the disk and stretching a cylindrical tube from one to the other—see figure 59. The number of handles on the disk is called the *genus* of the original surface; like the Euler characteristic, genus is a topological invariant of the surface. Using Seifert's construction, it is therefore possible to associate with any knot a certain natural number, namely, the genus of the Seifert surface for that knot. This number is called the *genus of the knot*. (Actually, there is a slight complication, in that Seifert's method can give different surfaces—with different genuses—starting from the same knot. So what you do is take the smallest genus number that arises from the knot in this way.)

Like the knot group, the genus has excellent potential as a way of determining whether a given knot is really knotted or not. The genus of the trivial knot is clearly zero, since the circle bounds a disk with no handles. Moreover, though not so obvious, the trivial knot is the only one with genus 0. So to show that a knot is really a knot, all you need to do is show that its genus is nonzero. But how do you go about calculating the genus of a knot—given, say, a diagram of the knot (and this is what you are usually presented with)? A possible method was suggested in 1962 by Wolfgang Haken (of four-color-theorem fame—see chapter 7), and more recently, in 1978, Geoffrey Hemion used Haken's results to construct an algorithm that always decides whether two given knot diagrams represent equivalent knots. Unfortunately, the algorithm is far too inefficient to be of much practical use (see chapter 11 for a discussion of the efficiency of algorithms), but it does show that the problem of distinguishing between knots is, in principle, one that can be decided in a mechanical fashion.

Incidentally, although Seifert's construction was developed geometrically rather than algebraically, in 1978 Charles Feustel and Wilbur Whitten showed how to obtain the genus from the knot group, thereby emphasizing once again the fundamental nature of the knot group.

By this time, you are no doubt beginning to reel at the amount of heavy mathematical machinery that has been brought into play in order to distinguish among knots. Isn't there a simpler way? you might ask. Well, if you are prepared to accept an occasional failure to distinguish among nonequivalent knots, there is indeed a simpler way. It is provided by the so-called knot polynomials, the simplest of which are the *Alexander polynomials,* discovered by J. W. Alexander in 1928. Although it may be derived from the knot group, the Alexander polynomial of a knot can be obtained directly from its diagram. For the trivial knot, the Alexander polynomial is the number 1. For the trefoil (figure 53(c)), it is

$$x^2 - x + 1,$$

and for the four-knot (figure 53(d)), it is

$$x^2 - 3x + 1.$$

Since these two polynomials are obviously not the same, it follows that the trefoil and the four-knot are not equivalent (and that neither of them is trivial). The Alexander polynomial of a knot is a very useful invariant: it is adequate to distinguish among all prime knots with crossing number up to 8 and all but six pairs with crossing number 9.

But one pair of knots that the Alexander polynomials cannot tell apart are the reef knot and the granny knot. These both have crossing number 6, but they are not prime, each being composed of two trefoils. The difference between them is whether the trefoils are left- or right-handed (see figure 55). Both knots have the Alexander polynomial

$$(x^2 - x + 1)^2.$$

(It is no accident that this is just the square of the Alexander polynomial of the trefoil. The Alexander polynomial of any composite knot is just the product of the Alexander polynomials of the constituent knots.) But don't blame the Alexander polynomial for this failure. As mentioned

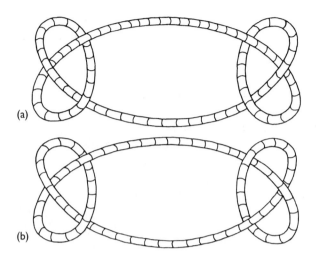

FIGURE 55   (a) A reef knot and (b) a granny knot. The difference between the two is the way the trefoil on the right winds around in the two knots. The knot group cannot distinguish between these two orientations.

earlier, the much more powerful knot group cannot distinguish between them either. The problem is in differentiating between a left-handed trefoil and a right-handed trefoil. The knot group cannot do this, at least not on its own. By "enhancing" the knot group in a suitable fashion, however, it is possible to obtain a knot invariant that will do the job, but once again the solution involves still more abstract machinery.

So is it the case that what Boy Scouts can do with ease has mathematicians searching through their toolboxes of highly sophisticated, abstract apparatus? Until August 1984, the answer was yes. But then things brightened up considerably when a New Zealand mathematician by the name of Vaughan Jones discovered a new class of polynomial invariant for knots, the *Jones polynomials*.

Besides being significant in itself, the discovery of the Jones polynomials led to a whole array of new polynomial invariants, resulting in a dramatic rise in research in knot theory, some of it spurred on by some exciting new applications in physics, which we shall discuss later in this chapter.

Whereas the Alexander polynomials have just one variable, the Jones polynomials have two. (They also involve negative powers of these variables, so you may at first balk at calling them polynomials.) The Jones polynomial for the right-handed trefoil is

$$-x^{-4} + x^{-2}y + x^{-2}y^{-1}$$

and the one for the left-handed trefoil is

$$-x^4 + x^2y + x^2y^{-1}.$$

For the four-knot, the Jones polynomial is

$$x^{-2} + x^2 - y - y^{-1} + 1.$$

How do you calculate these polynomials? Well, you obtain them from the knot diagram. Although not within the scope of a book like this, the procedure is sufficiently mechanical to be performed on a computer in an acceptably efficient manner.

Jones discovered the new polynomials while working on mathematical problems arising in physics. As Michael Atiyah has pointed out, the discovery was not just accidental; it revealed remarkable, deep connections between knot theory and mathematical physics. After Jones's discovery, Russian physicist Viktor Vassiliev found a whole new class of knot invariants based not on the topology of the individual knots but on the structure of the space of all closed curves. Knot theorists Joan Birman and Xiao-Song Lin at Columbia University in New York were able to show that Vassiliev's invariants were closely connected to the Jones polynomials, thereby strengthening the connection between knot theory and physics.

Another instance in which investigations in physics have led to developments in knot theory involve the so-called *unknotting number* of a knot. The unknotting number is an obvious way of classifying a knot, being defined as the least number of times the knot must be passed through itself in order to unknot it. Until recently, however, there was no general method for computing unknotting numbers, so the use of this invariant was strictly limited. When the first breakthrough came, the key idea again came from mathematical physics. In 1993, Peter Kronheimer and Tomasz Mrwoka proved a 40-year-old technical conjecture of John Milnor concerning the unknotting number of a certain class of knots. The conjecture said that if the knot can be drawn on a torus in such a way that it wraps through the hole p times and goes q times around the torus, its unknotting number is $(p-1)(q-1)/2$.

Kronheimer and Mrowka did not set out to solve Minor's problem. Rather, they were investigating surfaces in four-dimensional space, using some major new ideas introduced from mathematical physics by Simon Donaldson (see later in the chapter). Application of their work to knots came by viewing the deformation of a knot as occurring in time, when it sweeps out a surface in four-dimensional space–time.

Following Kronheimer and Mrowka's result, various mathematicians used similar techniques to obtain partial results regarding the unknotting number of general knots, leading knot theorists to speculate that at last they seemed to be making progress in understanding what Birman has described as "this mysterious number."

Meanwhile, as ideas from mathematical physics were proving useful in the pure mathematical study of knots, the Jones polynomials were finding themselves increasingly tied up with new developments in physics itself. The *superstring theory* advocated by physicist Edward Witten and others posits the universe as being made up of tiny, knotted coils, whose knot structure determines the structure of matter. Those new developments form the focus of the later parts of this chapter.

Do the various new invariants distinguish among all nonequivalent knots? No, they do not. For that you must keep on searching. In the theory of knots, much is still to be unraveled, but for us it is time to move on to something else.

Scratching the Surface

The central thrust in the development of topology since the mid-1950s has undoubtedly been in the study of *manifolds*. Loosely speaking (and a proper definition will be given later), a *manifold* is a generalization of the notion of a surface to any number of dimensions. So the simplest kinds of manifold are the one-dimensional ones, which are just curves (with the real line, $\mathbb{R}$, a special case) and the two-dimensional ones, the surfaces (with the flat two-dimensional plane, $\mathbb{R}^2$, a special case). Two-dimensional manifolds have the advantage that it is possible to draw pictures of them (or, even better, to make physical models of them). Unfortunately for the readers of this book, almost all the work on manifolds since the turn of the century has concerned manifolds of three or more dimensions, and such manifolds cannot be properly illustrated. (Diagrams in books dealing with higher-dimensional manifolds are always meant for the expert and require great care in their interpretation.) Of course, if the dimension is not too high, say three or four, then it is possible to illustrate simple manifolds using projections or cross sections. But again, these usually require some accompanying explanation to be understood. For instance, without the caption, would you have recognized the four-dimensional object that figure 56 is intended to illustrate? It shows two projections of a *hypercube*, a four-dimensional analogue of a cube. Just as a three-dimensional object can be illustrated on a two-dimensional page by means of a projection, so too—in principle—can a four-dimensional object be depicted by projecting it down to a three-dimensional object in space. By further projecting this three-dimensional object onto a page, it is possible to get an illustration of the original four-dimensional object, and this is what you see in figure 56. Of course, some considerable mental effort is required to interpret such

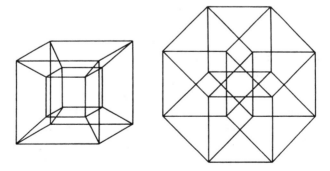

FIGURE 56   The hypercube, a four-dimensional figure having eight equal-sized cubes as its "faces." Both illustrations are flat projections of a (three-dimensional) "projection" of a hypercube into three-dimensional space.

a projection. With flat projections of three-dimensional objects, the human mind is quite good at doing this. Indeed, some of the disturbing art of the Dutch painter M. C. Escher achieves its effect by exploiting this facility in order to create "impossible" figures like the one shown in figure 57. Interpreting a projection of a four-dimensional object is more difficult—what does a hypercube "look like"? Well, just as an ordinary cube has six faces, each of which is a square and all of which are the same size, so a hypercube has eight "faces," each of which is a cube and all of which are the same size. If you look carefully at the left-hand diagram in figure 56 (and try to visualize the three-dimensional projection that it depicts), you will see that it shows eight cubes: a large outer one, a smaller inner one, and six that are distorted into the form of truncated pyramids. The distortions in size and shape all are features of the initial projection from four down to three dimensions. In (four-dimensional) "reality," all eight cubic "faces" are the same size, and the hypercube is what lies "between" them.

Besides projections, another way of trying to visualize higher-dimensional objects is with a succession of cross sections. For instance, what four-dimensional object gives the sequence of three-dimensional cross sections shown in figure 58? Well, successive two-dimensional cross sections of a sphere form a sequence of circles, starting from very small ones, growing to a maximum, then growing smaller again. (This is familiar to anyone who has ever sliced an apple.) In the same way, successive cross sections of a four-dimensional *hypersphere* form a sequence of spheres, as shown in figure 58. (Question: What does a sequence of cross sections of a hypercube look like?)

Such illustrative devices can give you, at best, a vague idea of what a four-dimensional object "looks like." They are no use whatsoever in trying to appreciate objects of five or more dimensions. But they do indicate how the progression from lower to higher dimensions proceeds in a step-by-step fashion of analogy. For instance, the bounding "faces" of a two-dimensional square are four equal-sized one-dimensional straight lines; the bounding faces of a three-dimensional cube are six equal-sized two-dimensional squares; the bounding "faces" of a four-dimensional hypercube are eight equal-sized three-dimensional cubes; and so on.

But beware! All is not as might at first appear. In three-dimensional space there are, as the ancient Greek geometers knew, just five regular polyhedra: the tetrahedron, cube, octahedron, dodecahedron, and icosahedron. In more than three dimensions, the analogue of a polyhe-

FIGURE 57   The art of M. C. Escher: *Ascending and Descending* (1960). The human facility for interpreting a three-dimensional object from its two-dimensional projection is exploited to achieve an impossible figure.

dron is called a *polytope*. (This concept is also mentioned in chapter 11, along with an application in the real, three-dimensional world.) The "faces" of a four-dimensional polytope are three-dimensional polyhedra. For a regular polytope, these "facing" polyhedra must themselves be regular, and the arrangement of the "faces" must be the same at each ver-

FIGURE 58   Cross sections of a four-dimensional hypersphere. This is what you would obtain if you were able to slice up a four-dimensional apple: a sequence of spherical "slices" that grows to a maximum size and then decreases to nothing again.

tex. It turns out that there are just six regular four-dimensional polytopes: the *simplex,* having as "faces" five tetrahedra; the *hypercube,* with eight cubes as "faces"; the *16-cell,* bounded by 16 tetrahedra; the *24-cell* with 24 octahedra; the *120-cell* with 120 dodecahedra; and the *600-cell,* with 600 "faces," each a tetrahedron. So things get a little more complicated as you move from three to four dimensions. But then something curious happens. For any number of dimensions greater than four, there are only three regular polytopes, analogous to the tetrahedron, cube, and octahedron. So what is going on? Why should things suddenly become simpler (and constant) beyond four dimensions? Although no one really knows the answer to this question, the phenomenon is not restricted to regular polytopes. In many other respects also, spaces of five or more dimensions are much easier to deal with than three- and four-dimensional space.

But despite the fact that some new factors come into play in higher dimensions, looking at what goes on for two-dimensional manifolds (surfaces) still offers a reasonable idea of the kinds of problems and methods involved in manifold theory, and consequently we should take a closer look at the topological theory of surfaces.

The classification of all two-dimensional manifolds, mentioned earlier, was one of the great triumphs of nineteenth-century topology. Only two invariants are required to distinguish among all closed surfaces: orientability and the Euler characteristic. How can this classification be achieved? Modern proofs usually proceed in two stages. First, it is shown that every closed surface can be topologically deformed into one of two standard forms. Then all that remains to be done is to show that the two invariants, orientability and the Euler characteristic, are enough to distinguish among all standard surfaces.

The *standard orientable surface of genus n* consists of a sphere to which are attached n handles. To attach a handle to a surface, you cut two holes in the surface and then sew in a cylindrical tube to join the two holes together (see figure 59). A sphere with any number of handles is an orientable surface. The Euler characteristic of a sphere with n handles is $2 - 2n$. This is not difficult to prove—the trick is to start with a network on a sphere (for which $V - E + F = 2$) and then add handles to it. If you do this carefully enough, you will see that each time a handle is added, the Euler characteristic decreases by 2.*

By means of a process consisting of cutting, pulling apart, and reassembly, known—for obvious reasons—as *surgery*, it is not at all difficult to deform any given orientable surface into a standard orientable surface of some genus. For instance, a torus gives one of genus 1; a dou-

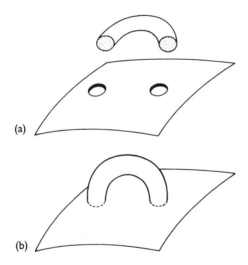

(a)

(b)

FIGURE 59   Handles. To attach a "handle" to a surface, cut two holes, as in (a), and sew a cylindrical tube to the two edges to connect them together, as in (b). Using a technique known as surgery, it is possible to show that every orientable closed surface is topologically equivalent to a sphere to which are attached a certain number of handles. This gives a "standard form" for orientable closed surfaces.

---

*Full details of this procedure, and indeed the entire classification proof, can be found in *Concepts of Modern Mathematics*, by Ian Stewart (Dover, 1995), chap. 12.

ble torus gives one of genus 2; and so on. Since the genus and the Euler characteristic for the standard orientable surfaces are related (by the preceding expression $2 - 2n$), this shows how the Euler characteristic serves to classify all orientable surfaces.

The *standard nonorientable surface of genus n* is obtained by adding *n* *crosscaps* to a sphere. To add a cross-cap to a surface, you cut a hole in the surface and sew a Möbius band across it, edge to edge. In three-dimensional space, you can do this only if you allow the Möbius band to intersect itself (see figure 60). Because a Möbius band enables clockwise and counterclockwise directions to be interchanged, a surface with a cross-cap is nonorientable. The Euler characteristic of a sphere with *n* cross-caps is $2 - n$. Again, this can be established by starting with a net-

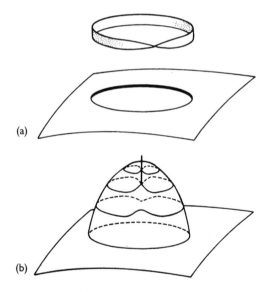

FIGURE 60   The cross-cap. To attach a cross-cap to a surface, cut a hole, as in (a), and sew a Möbius band across it edge to edge. In three-dimensional space, this can be visualized only if the Möbius band is allowed to intersect itself, as in (b). The resulting sewn-in piece is the cross-cap. By means of topological surgery, it is possible to show that any nonorientable closed surface is topologically equivalent to a sphere with a certain number of cross-caps. This gives a "standard form" for nonorientable closed surfaces.

work on a sphere and observing that each addition of a cross-cap decreases the quantity $V - E + F$ by 1. (This is fully explained in Stewart's book, cited in the footnote on page 249.)

By using surgery, it is possible (and not particularly difficult) to deform any nonorientable surface into a standard nonorientable surface of some genus. For instance, the geometer's *projective plane* (which, despite its name, is a closed surface) is transformed into a standard nonorientable surface of genus 1, and the Klein bottle is transformed into one of genus 2. Since the Euler characteristic of a standard nonorientable surface is related to the genus (by the expression $2 - n$), this standardization procedure establishes the classification by the Euler characteristic of all nonorientable surfaces as well. Surfaces with edges are dealt with by allowing the standard surfaces to have holes in them.

With this knowledge of surface topology as a jumping-off point, we can go on to see what has been happening in higher dimensions in recent years. We start by looking at one of the simplest types of manifold: the n-dimensional sphere, for n equal to 2, 3, 4, and beyond. It is with these manifolds that the most famous problem in topology is concerned.

The Poincaré Conjecture

The simplest of all closed two-dimensional surfaces is the sphere, the surface that acted as the starting point for the classification process described earlier. The n-dimensional analogue of a sphere is known as an *n-sphere.* (So the ordinary sphere is a 2-sphere.) Just as the 2-sphere is the surface of a three-dimensional solid ball, so the n-sphere is the "surface" of an $(n + 1)$-dimensional "solid ball." At the beginning of this century, when the French mathematician Henri Poincaré began to investigate higher-dimensional manifolds (and thereby essentially started the subject of manifold topology as it is understood today), he paid particular attention to the n-spheres. They should, after all, be very special, just as the 2-sphere is special among two-dimensional manifolds. In 1904, unable to prove what he thought was an extremely reasonable assertion about n-spheres, Poincaré formulated the assertion as a conjecture that was destined to become the most famous problem in the field. Like all good conjectures, it is both fundamental and easy to state.

Suppose you draw a closed loop on a 2-sphere. Then it is possible to

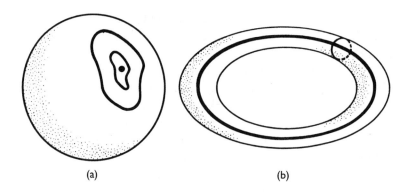

(a)   (b)

FIGURE 61   The Poincaré conjecture. If a closed loop is drawn on a sphere, it is possible to shrink that loop to a point without its leaving the surface, as in (a). On a torus, this is not always possible. If the loop is drawn in either of the two ways indicated in (b), it cannot be shrunk to a point without leaving the surface. For closed surfaces, this property of being able to shrink any closed loop to a point is characteristic only of the sphere; no other closed surface has this property. The Poincaré conjecture says that an analogous result holds for all higher dimensions. For instance, the only three-dimensional closed manifold with the loop-shrinking property is the three-dimensional hypersphere.

shrink the loop to a point without the loop's leaving the surface (see figure 61(a)). Moreover, the sphere is the only closed surface for which this is possible. If you were to draw your loop on a torus, for example, in either of the ways indicated in figure 61(b), then it could not be shrunk to a point. Similarly (only now you must rely on abstract mathematics without the aid of pictures), if you take an n-sphere for any value of n greater than 2 and "draw" a loop on it, then that loop may also be shrunk to a point without leaving the n-sphere. But (and this is the big question) is the n-sphere the only n-dimensional closed manifold that has this property, as it is in two dimensions? The *Poincaré conjecture* says that the answer is yes. (Strictly speaking, Poincaré was interested only in three-dimensional manifolds.)

Despite a great deal of effort, the problem of proving (or disproving) the Poincaré conjecture resisted all attempts at solution until 1960, when the American mathematician Stephen Smale proved that the conjecture is valid for all dimensions from five upward. This result was sufficiently highly regarded to win Smale a Fields Medal for his work. It also provided an instance of the phenomenon mentioned earlier, that manifolds behave differently from dimension five upward: Smale's methods did not work for three or four dimensions. In fact, almost twenty years went by before the problem was solved for four dimensions, also by an American. In 1981, Michael Freedman built on Smale's ideas and work by Andrew Casson in order to prove the Poincaré conjecture for 4-spheres. Freedman's result followed as a special case of a much more general result on four-dimensional manifolds, of which I shall say more in the next section.

This left the three-dimensional problem, the one for which the conjecture was originally formulated. What is the current state of affairs? Despite a great deal of work by first-rate mathematicians, it remains unsolved. Eighty years after it was put forward, the Poincaré conjecture is still not ready to give up its status as *the* unsolved problem in topology.

The Theory of Manifolds

The abstract definition of an n-dimensional manifold is that it is an object with the property that if you look at any small part of it, what you see looks very much like ordinary (?) n-dimensional Euclidean space, $\mathbb{R}^n$. For example, according to this definition, the 2-sphere is a two-dimensional manifold. Any small part of the sphere does indeed look like $\mathbb{R}^2$—a fact that is apparent to us as we walk around on the surface of the Earth. Note that because an n-dimensional manifold must look like $\mathbb{R}^n$ on a *local* basis, it does not follow that the entire manifold looks like $\mathbb{R}^n$. It was the failure to appreciate this (for two-dimensional manifolds) that led to the belief that the Earth is flat. Locally it is (almost) flat; globally it is not.

The main interest in the study of manifolds lies in the natural way they arise in connection with problems in analysis and physics, and in these cases you get more than a bare manifold as just defined. Rather, you get a manifold on which it is possible to develop one of mathematics' most powerful techniques: the differential calculus (i.e., you can "do differentiation" on it). You are probably familiar with the way the differential calculus is developed on the manifold $\mathbb{R}$. This is the "ordinary" differential calculus of functions of one real variable, taught in schools. You may even have seen how to do it on the manifold $\mathbb{R}^2$ as well. And the same kind of technique allows you to develop a differential calculus for any of the higher-dimensional manifolds $\mathbb{R}^n$, for $n = 3, 4, 5$, and so on.

Since any n-dimensional manifold is locally like $\mathbb{R}^n$, you can, of course, use the methods of the differential calculus on that manifold *on a local basis*. But what about globally? For the sphere at least, it is possible to develop a differential calculus that covers the entire surface. The reason is that the transition from one localized area (which looks like $\mathbb{R}^2$) to the next is essentially smooth and trouble free. To put it another way, suppose you were to cover the entire surface of the sphere with lines of latitude and longitude. Then you would get a coordinate system that, on a local basis, looks just like the usual coordinate system of Cartesian geometry (which is what lies behind the development of the differential calculus on $\mathbb{R}^2$). If you use these coordinates to develop your calculus locally on the surface, then because you are using the very same coordinate lines all over the surface, there will be no conflict between what happens at one location and what goes on at another: all the transitions will be *smooth* ones.

So a natural and fundamental question is, for how many manifolds is it possible to develop a differential calculus covering the entire manifold, as can be done for the sphere? A manifold for which it is possible to develop a global theory of differentiation is called a *smooth* (or sometimes a *differentiable*) *manifold.* A coordinate system that covers the entire manifold and serves as a basis for the differentiation process (such as the lines of latitude and longitude on the sphere) is called a *differentiation structure.* (Actually things are a bit more complicated, but this is more or less the picture.) The basic question of which manifolds are smooth (or which manifolds can be given a differentiation structure) carries with it the no less interesting ancillary question of whether, given a smooth manifold, there is only one differentiation structure that can be given to the manifold or whether it can be done in more than one way (and if so, in how many ways?). Moreover, since physicists spend much of their time working with the calculus on various manifolds, the answers to these questions are of interest not just to topologists.

For manifolds of two or three dimensions, the answers to these questions were known by the mid-1950s: every two- or three-dimensional manifold is smooth, and no such manifold can be given two essentially different differentiation structures. It was, it seemed then, only a matter of time before the result would be extended to cover manifolds of all dimensions. But in 1956, John Milnor of the United States discovered, much to everyone's surprise, that the 7-sphere can be given 28 distinct differentiation structures, and soon afterward it was found that other higher-dimensional spheres could be given more than one differentiation structure. Clearly, there was a great deal of work to be done to see what was going on, and there was no shortage of able mathematicians ready to do that work. The period between 1956 and 1970 has been called a "golden era of manifold topology." (What it was, in fact, was a golden era of the study of manifolds of five or more dimensions, for once again the problem for four dimensions proved intractable by the methods available.) During this period, by using a mathematical concept called *homotopy*, topologists were able to obtain a fairly systematic classification of all manifolds of dimension greater than four, distinguishing in particular between the smooth and the nonsmooth ones.

But what about four dimensions? Would all manifolds be differentiable and allow only a single differentiation structure, as with lower dimensions? Or would there be a whole range of possibilities requiring classification, as with higher dimensions? The answer finally came in 1981. Besides resolving the Poincaré conjecture for four dimensions (see earlier), Freedman also established that there is a four-dimensional manifold that is not smooth. (Freedman's description of this manifold—which for technical reasons is known as $\mathbb{E}_8$—is, like everything else in higher-dimensional topology, an algebraic one.) In fact, both the four-dimensional Poincaré conjecture and the nonsmooth four-dimensional manifold result followed from a single, very general (and totally unexpected) result that Freedman obtained, which showed that just two "elementary" pieces of information are all that is required to classify any four-dimensional manifold. (These pieces of information are not so "elementary" that they can be explained here, however.)

But that was not the end of the story. There was still another, equally dramatic surprise in store for the topologists, and it was not long in coming. And this surprise was one that hits right at the heart of the very physical universe we live in.

Unexpected results concerning manifolds can always be explained away on the grounds that one is, after all, dealing with some pretty abstract notions that can be at best only partly appreciated. Even two-dimensional manifolds can be pretty fancy objects, all but defying the imagination. But the same surely cannot be said of the "concrete" manifolds $\mathbb{R}$, $\mathbb{R}^2$, $\mathbb{R}^3$, and so on. After all, $\mathbb{R}^3$ is the physical space we live in (isn't it?), and $\mathbb{R}^4$ is the space–time continuum. And indeed these "concrete" manifolds do exhibit almost exemplary behavior. For a start, they are all smooth. Moreover, for each n, there is just one way of assigning a differentiation structure to $\mathbb{R}^n$—except, that is, for the one case $n = 4$.

For some peculiar reason, mathematicians had been unable to find a proof of the uniqueness of a differentiation structure that worked for $\mathbb{R}^4$. Every other $\mathbb{R}^n$, yes, but not $\mathbb{R}^4$. What made this lack of success all the more embarrassing was that this was the case of greatest interest to the physicists. Still, it could only be a matter of time before a proof was found. It was inconceivable that there could be a nonstandard way of doing differentiation on $\mathbb{R}^4$ . . . wasn't it?

But then the inconceivable turned out to be true. The bombshell was dropped in the summer of 1982. Combining Freedman's work (which is essentially algebraic) with a fairly hefty dose of analysis and differential geometry, Simon Donaldson, a 24-year-old student of Michael Atiyah at Oxford University, proved a result that implied the existence of a differentiation structure on $\mathbb{R}^4$ other than the usual one. In other words, the differentiation structure used by physicists and mathematicians the world over is not unique! (Indeed, subsequent work by Clifford Taubes showed that the usual differentiation structure on $\mathbb{R}^4$ is just one of infinitely many that may be given to this manifold!) This raises two intriguing questions. What is so special about four dimensions that makes this phenomenon occur only in this one case? And because there is more than one way of performing differentiation on $\mathbb{R}^4$, how do we know which is the "right" one as far as physics is concerned? With n-dimensional manifolds beginning to fall into order for all values of n other than 4, the case $n = 4$ has become curiouser and curiouser.

So, are physicists working with the right mathematics on $\mathbb{R}^4$? Well yes, they probably are. The infinitely many "exotic" differentiation structures on $\mathbb{R}^4$ that have been discovered all involve some very special "cooked-up" behavior that rules them out as far as our own physical universe is concerned. But their existence does indicate in a striking manner that there is something special about four-dimensional space.

QED

One of the more fascinating aspects of Donaldson's work was that it involved a reversal of the usual flow of ideas. Mathematicians and physicists are used to the notion that physics uses mathematics. But with Donaldson's work, the flow was in the opposite direction and heralded a new era in which geometry and physics became closely intertwined to a degree that no one had anticipated.

The notion that binds the two disciplines together is symmetry. Geometry can be regarded as the study of those properties of objects or figures that do not change when the objects or figures are moved around. An object is said to be symmetric if some motion leaves it looking exactly as it did in the first place, such as rotating a square about its central point through a right angle (see chapter 5).

Early in the twentieth century, physicists realized that many of their conservation laws arise from symmetries in the structure of the universe. For example, many physical properties are invariant under translation and rotation. The results of an experiment do not depend on where the laboratory is situated or the direction the equipment faces. This invariance implies the laws of conservation of momentum and angular momentum.

The German mathematician Emmy Noether proved that this holds in general, that every conservation law can be regarded as the result of some symmetry. Thus, every conservation law has an associated group—the corresponding symmetry group. For instance, the law of conservation of electric charge has an associated symmetry group, as do each of the nuclear physicist's laws of conservation of baryon number, of "strangeness," and of "spin." This connection makes it important for physicists to know about group theory, as understanding groups can help them come to grips with conservation. What has been happening in the years since Donaldson is that mathematicians wanting to come to grips with certain questions about groups have found it useful to understand conservation.

The mathematical task that led to this unexpected fusion of mathematics and physics is the classification problem for compact manifolds. (Compactness is a technical property of topological spaces. Intuitively, it means that if each point of the space gives you information about its immediate neighborhood, you can glean all the combined information from just finitely many of the points.) As described already under a slightly different guise, in the nineteenth century, mathematicians obtained a complete classification theorem for compact 2-manifolds: two compact 2-manifolds (i.e., two closed, orientable surfaces) are identical if, and only if, they can both be deformed into a sphere with the same number of handles. Thus the collection of spheres with handles provides a complete classification for compact 2-manifolds.

In the case of 3-manifolds, there is no analogous classification theorem, although the American mathematician William Thurston has produced a list of 3-manifolds from which, he conjectures, any compact 3-manifold can be assembled by stitching them together in certain ways. For 4-manifolds, there is not even a candidate list of classification manifolds.

Another way to view the classification problem is as a search for invariants that can be used to tell whether two given manifolds are really the same or are essentially different. What the post-Donaldson work in physics has done is give mathematicians new ways to obtain invariants for 4-manifolds.

The story begins in the 1920s with the introduction of quantum theory and, with it, a new version of Maxwell's electromagnetic theory, called *quantum electrodynamics,* or QED for short, thereby giving new meaning to the familiar abbreviation of the mathematicians' *quod erat demonstrandum.* As they tried to come to grips with the two electromagnetic theories, the classical and the newer quantum version, physicists began to see a common theme in the two: assigning a symmetry group to each point of four-dimensional space-time. In both cases, the group consists of the rotations of a circle, and the resulting theory is called *abelian gauge theory.* ("Abelian" because the group is abelian, and "gauge" because the physical motivation is of moving a measuring device, or gauge, around the space.)

Initially motivated by pure speculation, in the 1950s physicists began to investigate what would happen if they changed the group, say to the symmetries of a sphere or of other, higher-dimensional objects. The resulting theory was called *nonabelian gauge theory.* Central to the theory, and analogous to Maxwell's equations, are the Yang-Mills equations, proposed in 1954 by Chen-Ning Yang and Robert Mills. Just as Maxwell's equations describe an electromagnetic field and are associated with an abelian group, these new equations also describe a certain field,

but in this case, the field provides a means of describing particle interactions, and the associated group is noncommutative.

Just as abelian gauge theory has both classical and quantum versions,
so too the Yang-Mills equations can be treated both classically and in a
quantum-theoretic fashion. Whereas physicists were forced to use the
quantum version, mathematicians began to investigate the classical version. With ordinary four-dimensional space–time as the underlying
manifold, they were able to solve a special case of the Yang-Mills equations (the so-called self-dual case), calling their solutions *instantons*. Although of some interest to physicists, the mathematics of instantons was
still very much a pure mathematician's game, of interest to a relatively
small number of mathematicians. It was Donaldson's application of instantons and the self-dual Yang-Mills equations to general four-dimensional manifolds that opened things up dramatically, starting with the
surprising discoveries about the possible differentiation structures on $\mathbb{R}^4$
mentioned earlier.

One important consequence of Donaldson's theory was that it provided a means of generating invariants for 4-manifolds. Suddenly,
mathematicians had a way to approach the classification problem for
4-manifolds. But the going was decidedly tough: a lot of work was required to coax (pummel?) Donaldson's theory into turning out the desired invariants.

Michael Atiyah was sure that the solution to the difficulty, and the
solution to many difficulties that physicists were facing, was essentially
to reunite mathematics and physics over the issue. Inspired by Atiyah's
suggestion, in 1988 Princeton physicist Edward Witten managed to interpret Donaldson's theory as a quantum Yang-Mills theory, but apart
from making Donaldson's work intelligible to the physics community,
little came from it. Until 1993, that is. That year, things changed dramatically.

Physicist Nathan Seiberg had been looking at *supersymmetry*, a theory introduced in the 1970s that posits an overarching symmetry between the two main classes of elementary particles, the fermions (which
include quarks and electrons) and the bosons (which include photons
and gluons). He had been able to develop a lot of technical apparatus to
handle the difficult problems of quantum gauge theories for the super-
symmetric case. In 1993, Seiberg and Witten collaborated on an investigation of the case related to Donaldson's theory. In what Witten sub-

sequently described as "one of the most surprising experiences of my life," the two of them were able to solve the problem, in essence finding a new pair of equations to replace the ones coming from Donaldson's theory.

The discovery had immediate consequences not only for physics but also for mathematics. Although his main interests were in the physics, Witten worked out the implications of the new discovery for 4-manifolds. He conjectured that the new equations would produce invariants identical to those coming from Donaldson's theory.

Witten's conjecture was taken up by Clifford Taubes and others, and pretty soon people realized that something significant had happened. It was not clear that the invariants the new equations give rise to were exactly the same as Donaldson's invariants (and this remains an open question). However, the invariants they produced were at least as good, but much, much easier to deal with. Using the new equations, mathematicians were able to redo in a few weeks, and with far less effort, what had taken years using the old methods and to solve problems they had not been able to crack before.

The underlying distinction between Donaldson's theory and the new Seiberg-Witten theory that is the key to the power of the latter is compactness. The space associated with Donaldson's theory is not compact; the Seiberg-Witten one is. Compactness makes all the difference. As Taubes remarked, "We were quite embarrassed by the things we were proving, because they were quite literally observations. We felt like we were taking candy from a baby."

Thus, the Witten-Seiberg equations have given physicists a completely new way to regard the universe we live in and have also led to new mathematical results, as mathematicians have used ideas from physics to obtain new insights into manifold theory. Such a fruitful merger of topology, geometry, and physics had not been anticipated and indicates that for both physics and topology, we are definitely in the middle of a new golden age.

Suggested Further Reading

For a more complete introduction to topology than can be given here, though at essentially the same level, see *Concepts of Modern Mathematics*, by Ian Stewart (Dover, 1995), chaps. 10–14.

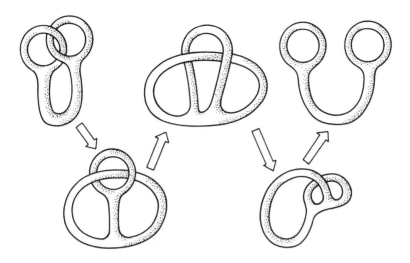

FIGURE 62   The solution to the ring puzzle (figure 48 on p. 226). The sequence shown indicates how the original linked ring configuration can be transformed into an unlinked ring figure.

Another good introduction is provided by *Knots and Surfaces*, by N. D. Gilbert and T. Porter (Oxford University Press, 1994), which, as the title suggests, includes material on both knots and surfaces.

A more substantial introduction to knot theory is provided by the book *Knot Theory*, by Charles Livingston (Mathematical Association of America, Carus Monographs, 1993). At a similar level, there is *The Knot Book*, by Colin Adams (W. H. Freeman, 1994).

For a readable overview of the tabulation of all prime knots with 16 or fewer crossings, see the article "The First 1,701,936 Knots," by Jim Hoste, Morwen Thistlethwaite, and Jeff Weeks, in the magazine *Mathematical Intelligencer* 20, no. 4, Fall 1998, pp. 33–48. Any reader who wants to know more would perhaps best be advised to seek out a friendly topologist to help, rather than trying to tackle any book unaided.

Other than at the level of, say, Stewart's book mentioned earlier, manifold theory tends to be fairly impenetrable to the outsider.

If you really *must* know more, why not go right to one of the main sources, with *The Geometry of Four-Manifolds*, by S. K. Donaldson and P. B. Kronheimer (Oxford University Press, 1990).

10

Fermat's Last Theorem

The Most Famous Problem in Mathematics

In October 1994, a British mathematician named Andrew Wiles announced that he had solved a problem that had resisted all attempts at a solution for more than three hundred years, a problem that had grown to become the most famous unsolved problem in mathematics: *Fermat's last theorem.* With the subsequent agreement among mathematicians that his proof is correct, Wiles brought to an end not only a three-hundred-year saga but also a nine-month period of uncertainty that had begun when a flaw had been found in an earlier proof he had announced in June 1993.

The fame of the problem was not restricted to the mathematical world—there can scarcely be an educated person anywhere who has not heard of Fermat's last theorem. And yet the origin of the problem amounts to no more than a scribbled note in the margin of a book.

When Pierre de Fermat died on January 12, 1665, he was one of the most famous mathematicians in Europe. Although his name is nowadays invariably associated with number theory, much of his work in this area was so far ahead of his time that to his contemporaries he was better known for his research in coordinate geometry (which he invented independently of Descartes), infinitesimal calculus (which Newton and Leibniz brought to fruition), and probability theory (which was essentially founded by Fermat and Pascal). And yet for all that, Fermat was not a mathematician by profession but a lawyer and magistrate attached to the provincial parliament in Toulouse, a position that he attained in 1631 when he was 30 years old.

It was after Fermat had taken up his jurist's post that he began to devote his leisure time to mathematics—a subject in which he had had no formal training but for which he rapidly developed a great affinity. What he did not develop an affinity for was preparing his work for publication. With a few exceptions, he published virtually nothing throughout his mathematical career. But Fermat did keep up a copious correspondence with his contemporaries—contemporaries who were among the greatest mathematicians alive. In a world peopled by such giants of mathematics as Girard Desargues, René Descartes, Blaise Pascal, John Wallis, and Jacques Bernoulli, this Frenchman for whom mathematics was a hobby could count himself the equal of any: Pierre de Fermat, the "Prince of Amateurs."

The path leading to the formulation of the famous last theorem is a long and interesting one. When Constantinople fell to the Turks in 1453, the Byzantine scholars fled to the West, bringing with them the ancient manuscripts of Greek learning. Among them was a copy of what had survived of Diophantus' *Arithmetica*. This work remained safe but largely unread until 1621, when Claude Bachet published a new edition of the original Greek text, along with a Latin translation containing notes and comments. This brought the book to the attention of the European mathematicians, and it seems that it was through reading the *Arithmetica* that Fermat first became interested in number theory.

The *Arithmetica,* written sometime in the third century A.D., was Diophantus' principal work and was one of the first books on algebra to be written. The greater part of the treatise concerns the solution in rational numbers of equations in two or more variables having integer coefficients. When working on such problems, present-day mathematicians usually restrict themselves to finding integer solutions. This often amounts to the same thing. For example, for a linear equation in three variables such as

$$2X + 3Y + 4Z = 0,$$

the rational solution $X = \frac{1}{4}$, $Y = \frac{1}{10}$, $Z = -\frac{1}{5}$ can be converted to the integer solution $X = 5$, $Y = 2$, $Z = -4$ by multiplying through by 20, the least common multiple of 4, 10, and 5. A similar procedure can be used in many other cases to convert a rational solution to one consisting entirely of integers. Certainly this is true of all the equations considered in this chapter, and so we too will consider only integer solutions.

As he worked through his copy of the Bachet edition of the *Arithmetica,* Fermat was in the habit of making brief notes in the margin. Five years after Fermat's death, when his son Samuel began to collect his father's notes and letters for publication, he came across the annotated copy of the *Arithmetica* and decided to publish a new edition of the book, including Fermat's marginal notes as an appendix. The second of these forty-eight *Observations on Diophantus* (as Samuel called them) had been written by Fermat in the margin next to Diophantus' problem 8 in book 2, which asks, "Given a number that is a square, write it as a sum of two other squares." Fermat's note said (in Latin):

> On the other hand, it is impossible for a cube to be written as a sum of two cubes or a fourth power to be written as a sum of two fourth powers or, in general, for any number that is a power greater than the second to be written as a sum of two like powers. I have a truly marvelous demonstration of this proposition that this margin is too narrow to contain.

In algebraic terms, Diophantus' problem asks for rational numbers x, y, and z satisfying the equation

$$x^2 + y^2 = z^2.$$

This turns out to be a fairly easy task. What Fermat's marginal comment asserts is that if n is a natural number greater than 2, then the equation

$$x^n + y^n = z^n$$

has *no* rational solutions. (As mentioned in chapter 3, in Diophantus' time—and indeed to some extent in Fermat's time—0 was not considered to be a number, so the trivial solutions you get by putting one of the variables equal to 0 are excluded here. The problem concerns positive rational solutions only.)

Note that by means of a simple argument like the one given a moment ago, it does not make any real difference if we amend both Diophantus' problem and Fermat's claim to refer to integer solutions (in fact, positive integer solutions) rather than rational solutions, since any rational solution leads at once to an integer solution (and conversely, any integer solution is clearly a rational solution). Thus we can take Fermat's last theorem (as his marginal claim is called) to be the assertion that for any natural number n greater than 2, the equation

$$x^n + y^n = z^n$$

has no positive integer solutions.

But why is it called "Fermat's last theorem"? The origin of this name is somewhat obscure. Although it is not known for certain when Fermat made his famous marginal note, it seems likely that it was during the period when he was first studying Diophantus' book, sometime in the 1630s. But this was at the start of his mathematical career, so the theorem was surely not his last one. Much more likely, the name stems from the fact that, of all the many statements of theorems Fermat left behind after his death, this is the last one that remained to be proved.

That might explain the use of the word *last*, but what about *theorem*? Did Fermat really have the "truly marvelous demonstration" that he claimed? Although this is a possibility, it is more likely that he was mistaken—and indeed, that he himself later realized his error. His other theorems are stated and restated in letters and challenge problems to other mathematicians, and the two special cases $x^3 + y^3 = z^3$ and $x^4 + y^4 = z^4$ of the last theorem are also stated elsewhere, whereas the only mention of the full last theorem is his one brief marginal note. Very likely Fermat saw how to prove it for $n = 4$, and possibly also for $n = 3$, and thought his arguments could be generalized to cover every other integer n, but subsequently discovered that this was not so. But because his marginal notes were never intended for publication, he would have had no need to go back and make any alteration. Indeed, Fermat may well have forgotten all about his earlier entry!

And yet in the popular mind, the belief persists that Fermat really did have a proof. It is, after all, a wonderful story: a seventeenth-century amateur proving a result that was to defeat the efforts of professional mathematicians for the next 350 years. The fact that the problem was so easy to state made the story even better.

But whether or not Fermat had a proof, the fact remains that for more than 350 years, no one else was able to solve this tantalizingly simple looking problem one way or the other. It was not for lack of trying. Many great mathematicians spent years of their lives wrestling with it, and work on the problem led to the development of totally new areas of mathematics. Entire books have been written about it (some of which are listed at the end of this chapter). In fact, the results that have been obtained as a consequence of trying to prove the last theorem, including the final proof, far outweigh the theorem itself in their significance for the rest of mathematics.

Indeed, the "theorem" itself has virtually no known consequences of any significance. Its importance rests solely on two things: its fame and the very fact that for so long, no one was able to solve it. To see what all the fuss was about, we need to go back to ancient Greece and, in particular, the mathematics of Pythagoras and of Diophantus.

Pythagorean Triples

The problem in Diophantus' *Arithmetica* that led to the formulation of Fermat's last theorem is to find a method for the solution (among the rational numbers, though we shall restrict ourselves to integer solutions) of the equation

$$x^2 + y^2 = z^2.$$

Because of the obvious connection with Pythagoras' theorem, any three integers x, y, z that satisfy this equation are said to form a *Pythagorean triple*. For example, the numbers 3, 4, and 5 form a Pythagorean triple because

$$3^2 + 4^2 = 5^2.$$

Once you have a Pythagorean triple, you can obtain from it an infinite collection of other Pythagorean triples. Simply multiply the three numbers in your triple by any other number you like. For example, multiplying the triple 3, 4, 5 by 2 gives 6, 8, 10, which is a Pythagorean triple because

$$6^2 + 8^2 = 10^2,$$

Multiplying by 3 gives the Pythagorean triple 9, 12, 15. And so on. But in a real sense, there is only one *solution* here, namely, 3, 4, 5, with the others simply variations of it. The solution 5, 12, 13, however, is a different solution (which in turn gives rise to its own infinite family of solutions). What distinguishes the solutions 3, 4, 5 and 5, 12, 13 from the infinitudes of solutions arising from them by multiplication by a constant is that these original solutions have no common factors, that is, no number that divides each of 3, 4, and 5, and no number that divides 5, 12, and 13.

In general, if *a, b, c* is any Pythagorean triple, so is any multiple *ma, mb, mc*; conversely, if *u, v, w* is any Pythagorean triple and if *d* is a common factor of *u, v,* and *w,* then *u/d, v/d, w/d* is also a Pythagorean triple. To highlight the special nature of the basic triples such as 3, 4, 5 and 5, 12, 13, mathematicians call Pythagorean triples *x, y, z* that have no common factor (other than 1) *primitive* Pythagorean triples. Essentially, Diophantus' problem is concerned with finding a way of determining all primitive Pythagorean triples.

A fairly straightforward piece of mathematical reasoning leads to the following formula for generating all possible primitive Pythagorean triples *x, y, z*:

$$x = 2st, \quad y = s^2 - t^2, \quad z = s^2 + t^2,$$

where *s* and *t* are any natural numbers such that *s* is greater than *t*; *s* and *t* have no common factor; and one of *s* and *t* is even and the other odd. For example, $s = 2$, $t = 1$ gives the triple $x = 4$, $y = 3$, $z = 5$; $s = 3$, $t = 2$ gives $x = 12$, $y = 5$, $z = 13$; $s = 4$, $t = 1$ gives $x = 8$, $y = 15$, $z = 17$; and so on.

The complete solution to Diophantus' problem I just described appeared in Euclid's *Elements* (ca. 350–300 B.C.).

The Case *n* = 4

Now that we have disposed of Diophantus' problem, what about Fermat's last theorem itself? This asserts (in integer form) that for every natural number *n* greater than 2, the equation

$$x^n + y^n = z^n$$

has no (positive) integer solutions. Just how do you go about proving (or, rather, trying to prove) this kind of statement?

A sensible first step is to look at a few special cases, say $n = 3$, $n = 4$, and $n = 5$; if you can solve those, you might see how to prove the entire theorem. This seems to be how Fermat himself must have approached the matter. The only concrete evidence we have concerns his work on a problem closely related to the case $n = 4$. Indeed, this work is almost the only piece of Fermat's own mathematical reasoning that anyone besides himself has ever seen. It consists of yet another marginal note in the *Arithmetica*. Interestingly enough in view of the marginal note in which the last theorem was stated, this note ends with the words "The margin is too small to enable me to give the proof completely and with all detail."

Before I explain Fermat's argument (and how it yields the case $n = 4$ of the last theorem), you might like to ask yourself just how, in general terms, you would go about trying to deal with (say) the $n = 4$ problem. You could start by trying a few values for x, y, z to see if any of them satisfy the equation, namely,

$$x^4 + y^4 = z^4.$$

(Your expectation presumably would be that you would find no solution, as Fermat claims is the case.) After trying various values (without finding a solution), you might be tempted to write a computer program to carry out a more extensive and systematic search for solutions, say trying all values of x, y, z from 1 to 10,000 (though there are more efficient ways of going about it). But even after several hours of computing, you still would not have succeeded, and then the futility of such an approach would become all too apparent. For no matter how powerful your computer and how efficient your method, this strategy could never succeed in proving Fermat's assertion (in the case $n = 4$). The reason is that the last theorem (with $n = 4$) asserts that *no* triple can be a solution to $x^4 + y^4 = z^4$, which is an assertion about an infinite collection of triples, and no amount of computation will enable you to handle an infinity of cases. Such an approach might succeed in *dis*proving the last theorem, since the discovery of a single solution to the Fermat equation would do that, but it could never prove the theorem. To prove the last theorem or any one case of it, a more subtle, mathematical, approach is required.

As so often happens in mathematics, the best bet is to look for a proof by contradiction. That is, you want (for $n = 4$) to prove that there is no solution to the equation $x^4 + y^4 = z^4$. Accordingly, you begin by assuming that there is a solution, say X, Y, Z, and then, on the basis of this assumption, proceed (by means of a mathematical argument) to deduce a contradiction. Once you have obtained your contradiction, you will have achieved your goal, since contradictory conclusions can be obtained only from false assumptions—in this case, the assumption that a solution did exist.

The problem facing you is how to deduce a contradiction from your assumption. A particularly powerful method for statements that, like the last theorem, involve the natural numbers is the so-called method of infinite descent, a method invented by Fermat and (he claimed) used by him as the basis of all his proofs in number theory. An illustration of the method is the very proof that Fermat scribbled in the margin of the *Arithmetica*. It involves what are known as *Pythagorean triangles*. (You will see shortly how this method relates to the case $n = 4$ of the last theorem.)

For obvious reasons, a triangle is called *Pythagorean* if it is right angled and all three sides have integer length. (In other words, a Pythagorean triangle is one whose sides form a Pythagorean triple.) What Fermat proved is that the area of such a triangle can never be a square (i.e., the square of an integer). His argument runs as follows:

Suppose there were a Pythagorean triangle whose area were a square. Let x, y, z be the lengths of the sides of the triangle, with z the hypotenuse (see figure 63). Thus, by Pythagoras' theorem, x, y, z satisfy the identity

$$x^2 + y^2 = z^2.$$

Let the area of the triangle be u^2, where u is an integer. Using the formula that says that the area of a triangle is half the product of the base and the height, we find that

$$u^2 = \tfrac{1}{2}xy.$$

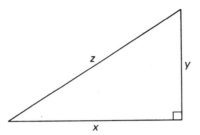

FIGURE 63 Fermat's result for Pythagorean triangles. According to Pythagoras' theorem, the sides of a right-angled triangle satisfy the equation

$$x^2 + y^2 = z^2.$$

The formula for the area of a triangle gives, in this case,

$$\text{Area} = \tfrac{1}{2}xy.$$

Fermat used the method of infinite descent to deduce a contradiction from the assumption that x, y, z all were integers and that the area of the triangle was the square of an integer.

By means of a rather ingenious argument, Fermat was then able to derive another set of positive integers X, Y, Z, and U such that

$$X^2 + Y^2 = Z^2, \quad U^2 = \tfrac{1}{2}XY, \quad Z < z.$$

(The argument is given in detail in chapter 1 of Harold Edwards's book on Fermat's last theorem—see the list at the end of this chapter.) The sought-after contradiction follows easily from this. The numbers X, Y, Z, U all have the properties that x, y, z, u have, so the same argument may be applied to them to obtain another four positive integers X_1, Y_1, Z_1, U_1 such that

$$X_1^2 + Y_1^2 = Z_1^2, \quad U_1^2 = \tfrac{1}{2}X_1 Y_1, \quad Z_1 < Z.$$

Similarly, there must exist another four positive integers X_2, Y_2, Z_2, U_2 such that

$$X_2^2 + Y_2^2 = Z_2^2, \quad U_2^2 = \tfrac{1}{2}X_2 Y_2, \quad Z_2 < Z_1.$$

And so on, ad infinitum. This process is known as *infinite descent* because the positive integers $z, Z, Z_1, Z_2, \ldots$ are getting smaller all the time (i.e., $z > Z > Z_1 > Z_2 > \ldots$). But this is the contradictory situation: there can be no infinite descending sequence of positive integers—eventually you get to 1 and then you must stop. The conclusion is that there cannot be a Pythagorean triangle whose area is the square of an integer.

Although there is no evidence that Fermat actually made the connection, it seems likely that he constructed the preceding proof in order to prove the case $n = 4$ of the last theorem. All that is required to deduce that case from the result on Pythagorean triangles is the following simple but ingenious trick.

Suppose that there is a solution to the equation $x^4 + y^4 = z^4$. In terms of this solution, set $a = y^4$, $b = 2x^2z^2$, $c = z^4 + x^4$, and $d = y^2xz$. Then, making repeated use of the familiar algebraic identity,

$$(r + s)^2 = r^2 + 2rs + s^2,$$

you get (check this for yourself)

$$
\begin{aligned}
a^2 + b^2 &= (z^4 - x^4)^2 + 4x^4z^4 \\
&= z^8 - 2x^4z^4 + x^8 + 4x^4z^4 \\
&= (z^4 + x^4)^2 \\
&= c^2.
\end{aligned}
$$

Also,

$$\tfrac{1}{2}ab = \tfrac{1}{2}y^4 2x^2z^2 = (y^2xz)^2 = d^2.$$

Thus $a^2 + b^2 = c^2$, and $\tfrac{1}{2}ab = d^2$. But this is precisely the situation that the preceding argument proved impossible. Hence the assumption that the equation $x^4 + y^4 = z^4$ has a solution must have been false. This completes the proof.

It follows immediately that the last theorem is true for n equal to any multiple of 4. For if the equation

$$x^{4k} + y^{4k} = z^{4k}$$

had a solution $x = a$, $y = b$, $z = c$, then a^k, b^k, c^k would be a solution of the equation $x^4 + y^4 = z^4$, which has just been proved to be impossible.

More generally, if the last theorem can be proved for any given exponent m, then it will be true for all multiples of m. Thus, because every integer greater than 2 is divisible either by a prime greater than 2 or by 4 (or both), in trying to prove the last theorem it is necessary to consider only those cases for which n is a prime greater than 2 (i.e., an *odd* prime) or else $n = 4$. Since the case $n = 4$ has just been disposed of, the problem reduces to the case in which n is an odd prime.

As mentioned earlier, there is no evidence that Fermat did prove the last theorem for $n = 4$. His proof of the result concerning Pythagorean triangles just outlined suggests that he probably did—certainly the step from there to the $n = 4$ case is one that was well within his capabilities. At any rate, most sources are content to credit him with the result. A similar cloud of uncertainty also lies over the proof of the next case to be solved, $n = 3$. Although this is almost universally credited to Leonhard Euler, the only published version of his proof likewise contains a "gap."

The Case n = 3

In a letter to Christian Goldbach dated August 4, 1753, Euler claimed that he had succeeded in proving Fermat's last theorem for $n = 3$. But he gave no proof in the letter. Nor indeed did he publish any proof until he included one in his book *A Complete Introduction to Algebra*, published in St Petersburg in 1770. Whether the proof he had in 1753 was correct is not known, but certainly the proof that appeared in 1770 contained a serious flaw. As it turned out, in the case $n = 3$ the flaw can be remedied, but in other cases a similar error proves to be insurmountable. Although Euler's argument is too long to give in detail, I shall outline it in general terms in order to explain just what the error was and why it was to prove so critical to later attempts to prove other cases of the last theorem.

As with Fermat's "proof" of the case $n = 4$, Euler used the method of infinite descent. Starting from the assumption that there was a solution x, y, z to the equation

$$x^3 + y^3 = z^3,$$

he was able to deduce that there was another solution X, Y, Z such that $Z < z$. The crux of the argument is to prove the proposition that if p and q are two numbers with no common factor and if $p^2 + 3q^2$ is a cube, then there must be numbers a and b such that $p = a^3 - 9ab^2$ and $q = 3a^2b - 3b^2$. This proposition is correct and can be proved by applying techniques found elsewhere in Euler's work. But in the proof of the last theorem that he published, Euler chose to employ a novel type of argument involving numbers of the form $a + b\sqrt{-3}$ (where a and b are integers), and this is where the error came in.

We can see why Euler found the numbers $a + b\sqrt{-3}$ useful if we expand the expression $(a + b\sqrt{-3})^3$. It is equal to

$$a^3 + 3a^2b\sqrt{-3} - 9ab^2 - 3b^3\sqrt{-3},$$

which can be rearranged as

$$(a^3 - 9ab^2) + (3a^2b - 3b^3)\sqrt{-3}.$$

So if $p = a^3 - 9ab^2$ and $q = 3a^2b - 3b^2$ (as stated in the conclusion of the proposition to be proved), then

$$p + q\sqrt{-3} = (a + b\sqrt{-3})^3.$$

Now, an assumption of the proposition is that $p^2 + 3q^2$ is a cube. This is equivalent to the statement that $(p + q\sqrt{-3})(p - q\sqrt{-3})$ is a cube, so the entire proposition can be rephrased as follows: if p and q have no common factor and if $(p + q\sqrt{-3})(p - q\sqrt{-3})$ is a cube (in the system of $\sqrt{-3}$ numbers), then $p + q\sqrt{-3}$ must be a cube—again, in the sense that $p + q\sqrt{-3} = (a + b\sqrt{-3})^3$ for some integers a, b.

To prove this reformulated version of the proposition, Euler reasoned as follows: The numbers $a + b\sqrt{-3}$ (for varying integers a, b) form a number system much like the integers. (See chapter 3 for a fairly full explanation of this issue.) Now, if m and n are two given integers with no common factor and if mn is a cube, then both m and n are cubes. By analogy, Euler argued that the same would be true for the $\sqrt{-3}$ system of numbers. As he correctly proved, the assumption that

p and q have no common factors implies that the $\sqrt{-3}$ numbers $p + q\sqrt{-3}$ and $p - q\sqrt{-3}$ have no common factors (among the $\sqrt{-3}$ numbers), and from that the desired conclusion follows at once.

The problem with this argument—and it is a serious one—is that reasoning by analogy with the integers is not valid. Just because the $a + b\sqrt{-3}$ number system resembles the integers in many ways (both systems form an integral domain—see chapter 3), it does not follow that this system has all the properties of the integers. (It does not.) As for the crucial property for Euler's proof, this holds for integers by virtue of the fundamental theorem of arithmetic, the unique factorization theorem, which says that every integer is a product of a unique collection of primes (and possibly $- 1$). Now, if you have read chapter 3, you will be aware that the $\sqrt{-3}$ number system does have this property. So Euler's conclusion is valid. But you will also know from chapter 3 that $\sqrt{-3}$ is one of only nine integer roots that does lead to the unique factorization property, so it is only by a stroke of luck that Euler's analogy argument did not lead to a false conclusion. Had he been trying instead to prove the case $n = 5$ of the last theorem, using the numbers $a + b\sqrt{-5}$, his method would have failed. As will be seen presently, the failure of unique factorization was to prove a rock on which many purported proofs were to founder.

Two More Cases: $n = 5$ and 7

In 1825, Peter Gustav Lejeune Dirichlet (who had just turned 20) and Adrien-Marie Legendre (who was past 70) proved the last theorem for the case $n = 5$. Their method was basically an extension of the one Euler had used for $n = 3$. The analogue of the critical equation

$$p + q\sqrt{-3} = (a + b\sqrt{-3})^3$$

was the identity

$$p + q\sqrt{5} = (a + b\sqrt{5})^5.$$

However, to prove that $p + q\sqrt{5}$ is a fifth power (for which argument by analogy with the integers is certainly not valid), they had to assume not only that $p^2 - 5q^2$ is a fifth power and p, q have no common factors (as with $n = 3$) but also that just one of p, q is even and that q is divisible by 5. They did not use (the nonexistent) unique factorization.

With the case $n = 5$ disposed of, the approach used by everyone so far began to show signs of strain as the demands made on the algebra became more and more severe. In 1832, having failed to get the method to work for $n = 7$, Dirichlet did succeed in proving the case $n = 14$ (a much weaker result, of course). When, in 1839, Gabriel Lamé finally did prove the case $n = 7$, he had to resort to some ingenious devices closely tied to the number 7 itself, and there seemed to be little hope of anyone being able to move on to the next case, $n = 11$, without adopting a radically new approach. It was Lamé himself who, in 1847, proposed just such a course of action.

The Cyclotomic Integers and Lamé's Announcement

Lamé's proposal was to try to prove the full last theorem by using a complex nth root of unity, that is, a complex number r for which $r^n = 1$ but $r^k \neq 1$ for any positive integer k less than n. (All this is for any odd prime n.) For any odd prime n (indeed for any odd number n), the number 1 has $n-1$ complex nth roots. For instance, for $n = 3$, the two complex cube roots of 1 are

$$-\frac{1}{2} + \frac{\sqrt{3}}{2}\,\mathrm{i}, \quad -\frac{1}{2} - \frac{\sqrt{3}}{2}\,\mathrm{i}.$$

(You can check this by cubing each of these complex numbers.)

The point of introducing such an r is this. The proofs of the cases $n = 3, 4, 5, 7$ that had been found up to that time all depended on some algebraic factorization such as, for $n = 3$,

$$x^3 + y^3 = (x + y)(x^2 - xy + y^2).$$

Lamé realized that the increasing difficulty for larger n was caused by the increasingly large degree of one of the factors in such a factorization. By introducing r, it is possible to factor $x^n + y^n$ completely into n factors, each of degree 1.

To obtain the factorization, note that the complex numbers 1, r, r^2, $\dots$, r^{n-1} are the roots of the complex equation

$$z^n - 1 = 0,$$

and so

$$z^n - 1 = (z - 1)(z - r)(z - r^2) \ldots (z - r^{n-1}).$$

If you now put $z = -x/y$ and multiply both sides of the equation by y^n, you get (since n is odd)

$$x^n + y^n = (x + y)(x + yr)(x + yr^2) \ldots (x + yr^{n-1}).$$

Each of the complex factors of $x^n + y^n$ in this expression is a special case of numbers of the general form

$$a_0 + a_1 r + a_2 r^2 + \ldots + a_{n-1} r^{n-1},$$

where $a_0, a_1, \ldots, a_{n-1}$ are integers. Numbers of this type—that is, numbers made up of integers and the powers of r—are nowadays known as *cyclotomic integers*. (All this is still relative to some fixed odd prime n.) Like the Gaussian integers or the numbers of the form $a + b\sqrt{-3}$ that appeared earlier, the cyclotomic integers produce a number system that to some extent resembles the ordinary integers. (They constitute a *ring*; see chapter 3 for the relevant definitions.)

On March 1, 1847, a highly excited Lamé stood up to address the members of the Paris Academy. He had, he said, at long last succeeded in proving Fermat's last theorem. His key idea was to work with (what are nowadays known as) the cyclotomic integers. This enabled him to carry out a proof by infinite descent, much like Euler's argument for the case $n = 3$. (Thus a crucial step in his argument was to show that if the factors $x + y, x + yr, \ldots, x + yr^{n-1}$ of $x^n + y^n$ have no common factors, then the equality of $x^n + y^n$ and z^n would imply that each of the factors $x + y, x + yr, \ldots, x + yr^{n-1}$ must be an nth power.)

Having described his purported proof, Lamé concluded by acknowledging that the idea for using the complex numbers in that way had been suggested to him by his colleague Joseph Liouville a few months before. It was Liouville himself who took the floor when Lamé sat down. Was Lamé really justified, he asked, in concluding that each factor of $x^n + y^n$ was an nth power if all that he had shown was that no two of these factors had a common divisor? The truth of such a step for the ordinary integers was, Liouville pointed out, dependent on the unique factorization theorem, and he knew of no such result for the cyclotomic integers.

Whether Liouville knew in advance of Euler's previous error on this point is not known. At any rate, his remarks struck at the very heart of Lamé's argument, and it was a much saddened and embarrassed Lamé who, after several weeks' valiant effort trying to rescue his proof, finally realized the enormity of his mistake. "If only you had been in Paris or I had been in Berlin, all of this would not have happened," he wrote to his friend Dirichlet in Berlin. In fact, Lamé's acute embarrassment would have been avoided had he only known of some work that had been published by one Ernst Eduard Kummer some three years earlier. (To be fair to Lamé, however, Kummer, for reasons known only to himself, had chosen the incredibly obscure journal *Gratulationschrift der Universität Breslau zur Jubelfeier der Universität Königsberg* as the resting place for his paper.)

Kummer's Work and Ideal Numbers

In his 1844 paper, Kummer had proved that unique factorization is usually false for the cyclotomic integers, a result that completely destroyed Lamé's purported proof of the last theorem. But by 1847, when Lamé and the rest of the mathematical world learned of these results, Kummer had developed an impressive new theory that showed that there was a way to modify the concept of unique factorization so that a reasonable "number theory" could still be obtained for the cyclotomic integers. The basis of his theory was the introduction into the arithmetic of the cyclotomic integers of what he called *ideal prime factors*—a step somewhat analogous to the introduction of the imaginary number i into the arithmetic of the ordinary integers. Using Kummer's *ideal numbers,* many of the consequences of unique factorization for the integers may be proved for the cyclotomic integers (and for other number systems such as $a + b\sqrt{-3}$ that arise in proving various cases of Fermat's last theorem).

Kummer's work marked the biggest single advance made in the study of Fermat's last theorem from its inception until a 1983 discovery (about which more presently). His 1847 results proved the last theorem for all prime exponents less than 37 (thereby establishing the result for all exponents less than 37) and for all prime exponents less than 100 with the exception of 37, 59, and 67. All this came just a few years after mathematicians had struggled with proofs for $n = 5$ and $n = 7$.

Moreover, Kummer's critical new concept of *ideal numbers* turned out to be an extremely powerful and wide-ranging one that led to a more general concept of an *ideal* and an entire branch of mathematics— *ideal theory*—whose rudiments are now taught as a matter of routine to all university mathematics students. Indeed, dramatic though the application of Kummer's ideal numbers to Fermat's last theorem was, it is the concept of an ideal itself that has proved to be the most important aspect of Kummer's work for the rest of mathematics.

In fact, just as the long-term significance of Kummer's work was not its consequence for the last theorem, neither did his work stem from an attempt to prove Fermat's claim. Like Gauss (see chapter 3), Kummer had been working on the problem of higher reciprocity laws to generalize the quadratic reciprocity law proved by Gauss, work that led to his proving in 1859 a powerful general result on the problem. That said, we should note that there is a close connection between Fermat's last theorem and the higher reciprocity laws.

Regular Primes

Kummer's work was particularly relevant to the last theorem because it provided an arithmetical condition that an odd prime exponent must satisfy for the last theorem to be true for that exponent. That is, if an odd prime p satisfies Kummer's condition, then the equation

$$x^p + y^p = z^p$$

has no solution. Nowadays, primes that satisfy Kummer's condition are known as *regular primes*. Of the primes less than 100, only 37, 59, and 67 fail to be regular, as Kummer himself verified in 1847.

What, then, is a regular prime? The concept is closely connected to that of *class number*, described in chapter 3. A *regular prime* is one that does not divide the class number of the associated cyclotomic number field. This uses some deep mathematical concepts, but there is an equivalent definition that is much simpler. Recall from chapter 3 that the irrational number e is a standard mathematical constant whose infinite decimal expansion begins 2.71828 . . . and that for any number t the value of e^t is given by the infinite sum

$$e^t = 1 + \frac{t}{1!} + \frac{t^2}{2!} + \frac{t^3}{3!} + \ldots .$$

The *Bernoulli numbers, B_k,* are defined as the coefficients in the infinite sum

$$\frac{t}{e^t - 1} = 1 + B_1 \frac{t}{1!} + B_2 \frac{t^2}{2!} + B_3 \frac{t^3}{3!} + \ldots .$$

The values of the Bernoulli numbers behave quite irregularly. B_k is zero for all odd k except $k = 1$, for which $B_1 = -\frac{1}{2}$. The first few values of B_k for even k are

$$B_2 = \frac{1}{6}, \quad B_4 = -\frac{1}{30}, \quad B_6 = \frac{1}{42}, \quad B_8 = -\frac{1}{30},$$

$$B_{10} = \frac{5}{66}, \quad B_{12} = -\frac{691}{2730}, \quad B_{14} = \frac{7}{6},$$

$$B_{16} = -\frac{3617}{510}.$$

In terms of the Bernoulli numbers, a prime p is regular if it does not divide the numerators of each of the numbers $B_2, B_4, \ldots, B_{p-3}$. (So p will be *irregular*—that is, will fail to be regular—if p does divide the numerator of at least one of the Bernoulli numbers $B_2, B_4, \ldots, B_{p-3}$.)

The definition of regularity in terms of Bernoulli numbers provided a way of checking the regularity of a given prime by computation, although using this definition directly is not very efficient, and in practice, various facts about the Bernoulli numbers are used to derive more manageable methods. (One problem facing anyone who sets out to check for regularity is that some very large numbers appear as numerators in Bernoulli numbers. For example,

$$B_{34} = \frac{2{,}577{,}687{,}858{,}367}{6}$$

is still manageable by hand, but the numerator of B_{220} has 250 digits!)

The first calculations to determine the regular and the irregular primes were carried out by Kummer himself, as I mentioned earlier. He went as far as 164, finding that the only irregular primes below this were 37, 59, 67, 101, 103, 131, 149, and 157. (In each case, Kummer had to show that the prime in question divided the numerator of an appropriate Bernoulli number. For example, 37 divides the numerator of B_{32}; 59 divides the numerator of B_{44}; and 157 divides the numerators of both B_{62} and B_{110}.)

Then during the 1930s, H. S. Vandiver used desk calculators (together with some new methods for checking regularity and irregularity) to test all the primes up to 617. In 1954, with the advent of electronic computers, Lehmer and Vandiver took the computation as far as 4001, and various people subsequently helped reach 30,000. At this point, in 1976, armed with both an IBM 360–65 and an IBM 370 computer, Samuel Wagstaff determined the regularity status of every prime below 125,000. By the early 1990s, computations of this kind had been carried as far as 4,000,000.

On the basis of all this numerical work, it seems that about 60 percent of all primes are regular. To be precise, for large N the observed ratio is

$$\frac{\text{Number of irregular primes less than } N}{\text{Number of primes less than } N} = 0.39.$$

A plausible though nonrigorous argument produced by Siegel in 1964 showed that this ratio "ought to be" $1 - 1/\sqrt{e}$. To two places of decimals, this works out to be (wait for it) 0.39.

And yet despite the apparent preponderance of regular primes, it is still not known for sure whether there are infinitely many of them. Surprisingly, in view of the numerical evidence, it is known that there are infinitely many irregular primes. This was proved by a mathematician called Jensen in 1915. So the seemingly larger set could turn out to be finite, even though the apparently smaller set is already known to be infinite.

But what did all this work tell us about Fermat's last theorem? Kummer's 1847 result showed that if p is a regular odd prime, then the last theorem is true for exponent p. But what happens when p is an irregular prime? Here Kummer's result did not help. It does not, of course, follow that the last theorem is false in this case. All that happens is that the specific argument Kummer used is no longer applicable. The result itself could still be true for some other reason.

Indeed, Kummer subsequently found conditions more inclusive than regularity, which still implied the last theorem for the exponent concerned and which could be verified by computation. Since these conditions were satisfied by the irregular primes 37, 59, and 67, he was thereby able to verify the last theorem for all exponents up to 100.

Over the years, various workers adopted the same overall approach of finding more general, computationally verifiable conditions on primes that implied the last theorem for the exponent concerned. By the early 1990s, with the aid of increasingly powerful computers, the last theorem had been verified in this way for all exponents up to 4,000,000 (the computational work mentioned a moment ago).

But such an approach could never yield the full theorem; for that, quite a different approach was required. One of the key ideas that did eventually lead to a proof was to reformulate the problem in geometric terms.

The Geometry of Fermat's Last Theorem

First, notice that finding *whole-number* solutions to an equation of the form

$$x^n + y^n = z^n$$

(including the case $n = 2$) is equivalent to finding *rational-number* solutions to the equation

$$x^n + y^n = 1.$$

By formulating Fermat's problem as one about rational solutions to these equations (which I shall henceforth refer to as *Fermat equations*), ideas of geometry and topology may be brought to bear. For instance, the equation

$$x^2 + y^2 = 1$$

is the equation of the circle of radius 1, having its center at the origin. Asking for rational-number solutions to this equation is equivalent to asking for points on the circle whose coordinates are both rational. Because the circle is such a simple mathematical object, it is easy to find such points, and indeed, a fairly straightforward geometric argument leads to the two formulas given earlier that generate all the Pythagorean triples.

However, the curves that correspond to the general equation $x^n + y^n = 1$, when n is greater than 2, seem no easier to analyze than the original equation. To make progress, additional structure is required, beyond the geometry of the curve.

First, the problem must be generalized by allowing *any* polynomial in two unknowns. Moreover, the unknowns x and y in the equation must be regarded as ranging not over real numbers but over complex numbers. Then, instead of giving rise to a *curve*, the equation determines a *surface*. (Figure 64 shows the surface that arises for the Fermat

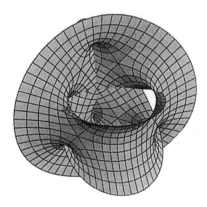

FIGURE 64   The surface $x^3 + y^3 = 1$. The surface is of genus 1 and has one hole through it.

equation with exponent $n = 3$.) In the topological terminology of chapter 9, the surface obtained in this manner is a closed, orientable surface. (Actually, not all equations give rise to a nice, smooth surface, but with some extra effort, it is possible to patch things up so that everything works out.) The crucial point about this step is that surfaces are nice, intuitive objects, about which there is a wealth of mathematical theory available. In particular, it is possible to use the techniques and results outlined in chapter 9, such as the fact that every closed, orientable, smooth surface is topologically equivalent to a sphere with a certain number of handles. The number of handles in this case is called the *genus of the surface,* so it is natural to call this number the "genus" of the original Fermat equation. For the Fermat equation with exponent n, the genus works out to be $(n - 1)(n - 2)/2$. For instance, the Fermat equation with exponent $n = 3$ has a surface of genus 1.

The problem of finding rational solutions to a polynomial equation in two unknowns turns out to be closely related to the genus of the equation. The larger the genus is, the more complicated is the geometry of the surface, and the harder it becomes to find rational points.

The simplest case is when the genus is 0, for example, the "Pythagorean" equation

$$x^2 + y^2 = k.$$

In this case, one of two results is possible. One possibility is that the equation has no rational points, which is what occurs for the equation

$$x^2 + y^2 = -1.$$

Alternatively, if there is any rational point, then it turns out that there are infinitely many rational solutions, and there is a formula that generates them all, as in the case of the Pythagorean triples.

The case of curves of genus 1 is much more complicated. Curves determined by an equation of genus 1 are called *elliptic curves,* since they arise in the course of computing the length of part of an ellipse. Roughly speaking, an elliptic curve is one having an equation of the form

$$y^2 = x^3 + Bx + D$$

where B and D are integers. Elliptic curves often consist of two disjoint pieces, as in the example illustrated in figure 65.

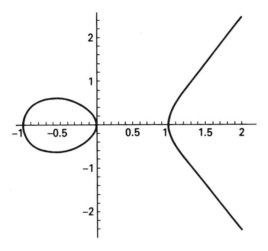

FIGURE 65   The elliptic curve $y^2 = x^3 - x$.

Elliptic curves have a number of nice properties that make them extremely useful in number theory. For example, some of the most powerful known methods for factoring large integers into primes (on a computer) are based on the theory of elliptic curves.

As in the case of curves of genus 0, an elliptic curve may have no rational points. But if there is a rational point, an interesting thing happens, as the English mathematician Lewis Mordell discovered in the early part of the century. Mordell showed that although the number of rational points may be either finite or infinite, there is always a *finite* set of rational points—they are called *generators*—such that all other rational points may be generated from them by a simple, explicit process. All that is required is some elementary algebra coupled with the drawing of lines that either are tangent to the curve or cut it at three points. Thus, even in the case in which there are infinitely many rational points, there is a means to handle them. (In fact, the rational points constitute a group.)

Of course, the genus 1 case is not in itself of particular interest if the goal is to prove Fermat's last theorem, which, with the exception of exponent $n = 3$, concerns equations of genus greater than 1. But as a result of his investigations, in 1922 Mordell made an observation of great potential relevance to Fermat's last theorem: no one had ever found an equation of genus greater than 1 that had infinitely many rational solutions. In particular, all the many equations that Diophantus had examined turned out to have genus 0 or 1. Mordell conjectured that this was not just an accident but that no equation of genus greater than 1 could have an infinite number of rational solutions.

In particular, Mordell's conjecture implied that for each value of the exponent *n* greater than 2, the Fermat equation

$$x^n + y^n = 1$$

could have, at most, a finite number of rational solutions. This meant that a proof of the Mordell conjecture, though not proving Fermat's last theorem, would mark a significant development.

The Mordell conjecture was finally proved in 1983 by a German mathematician named Gerd Faltings, an accomplishment for which Faltings was awarded the Fields Medal in 1986. It was by far the most significant advance on Fermat's last theorem since Kummer's work more than a century earlier.

The Proof of Fermat's Last Theorem

With the proof of the Mordell conjecture, it was known that for no exponent *n* greater than 2 does the equation

$$x^n + x^n = 1$$

have more than a finite number of rational solutions. However, there seemed to be no way to strengthen Faltings's proof to yield the stronger result claimed by Fermat, that there is *no* rational solution for any exponent *n* greater than 2.

And so it turned out. When the proof finally came, it followed quite a different path, a path that had its beginning back in 1955.

In that year, the Japanese mathematician Yutaka Taniyama suggested that there was a connection between elliptic curves and another well-understood (but not easily explained) class of curves, known as *modular curves*. Every elliptic curve and every modular curve has an associated "L-series," a complex function closely associated with prime numbers. (The

Riemann zeta function described in chapter 8 is an L-series.) In the early 1950s, Martin Eichler of the University of Basel noticed that the L-series for the elliptic curve $y^2 + y = x^3 - x^2$ was the same as the L-series for a particular modular curve. Taniyama's suggestion amounted to Eichler's observation being a special case of a general connection.

Taniyama's suspicion was made more precise in 1968 by André Weil, who showed how to determine the exact modular curve that should be connected to a given elliptic curve, and in 1971, Goro Shimura demonstrated that Weil's procedure worked for a very special class of equations. Taniyama's original suspicion became known as the *Shimura-Taniyama conjecture.*

At the time, no one suspected there would be any connection between this very abstract conjecture and Fermat's last theorem. But in 1986, a German mathematician named Gerhard Frey noticed that the two might in fact be very closely related.

Frey realized that if there are whole numbers a, b, c, n such that $c^n = a^n + b^n$, then it seemed unlikely that one could understand the elliptic curve given by the equation

$$y^2 = x(x - a^n)(x + b^n)$$

in the manner suggested by the Shimura-Taniyama conjecture. Later that same year, the American mathematician Kenneth Ribet proved conclusively that the existence of a counterexample to Fermat's last theorem would definitely lead to the existence of an elliptic curve that could not be modular and hence would contradict the Shimura-Taniyama conjecture. In other words, Ribet proved that the Shimura-Taniyama conjecture implied Fermat's last theorem.

Ribet's remarkable result was electrifying. The Shimura-Taniyama conjecture concerned geometric objects about which a great deal was known. Indeed, there was good reason to believe the conjecture. There also were some obvious—but daunting—ways to set about finding a proof. At last, mathematicians had a powerful framework with which to approach the last theorem. Among those who took up the challenge was the Princeton, New Jersey–based English mathematician Andrew Wiles.

Wiles had been fascinated with Fermat's last theorem since childhood, when he had attempted to solve the problem using high-school mathematics. Later, when he learned of Kummer's work, he tried again using the German's more sophisticated techniques. But knowing just how many mathematicians had tried and failed to solve the problem, he eventually gave up on the ancient problem and, as a graduate student in Cambridge, concentrated on more mainstream contemporary number theory, in particular the theory of elliptic curves, a subject attracting considerable interest at the time (but not thought to have any connection to Fermat's last theorem).

Then came Ribet's astonishing and completely unexpected result, and suddenly Wiles found that the area of number theory in which he had become an acknowledged world expert might provide the elusive key to the proof of Fermat's last theorem. For the next seven years, Wiles concentrated all his efforts on trying to find a way to prove the Shimura-Taniyama conjecture. By 1991, he felt sure he could prove not the entire Shimura-Taniyama conjecture, which applies to all elliptic curves, but a special case of the conjecture that applies to elliptic curves of a particular kind. In 1993, after two more years of work, he eventually succeeded in doing just that.

Believing that the class of elliptic curves for which his proof worked included those necessary to deduce the last theorem, in June 1993 Wiles announced that he had solved Fermat's last theorem.

He was wrong. By December of that year, he had to admit that his argument did not seem to work for the "right" elliptic curves. Although everyone agreed that his achievement was one of the most significant advances in number theory in the twentieth century, it appeared that Wiles was destined to follow in the footsteps of all those illustrious mathematicians of years gone by (including perhaps Fermat himself) who had dared rise to the challenge laid down in Fermat's marginal note. The last theorem remained unproved.

Several months of silence followed while Wiles retreated to his Princeton home to try to make his argument work. In October 1994 came the announcement that he had succeeded. His proof—a correct proof this time—was given in two papers. The first of these was a long one entitled "Modular Elliptic Curves and Fermat's Last Theorem," which contained the bulk of his argument. The second paper, entitled "Ring Theoretic Properties of Certain Hecke Algebras," was written jointly by Wiles and a former student, Richard Taylor, of Cambridge University, and provided a key step Wiles used in his proof. The two papers together constituted the May 1995 issue of the prestigious journal *Annals of Mathematics*.

With Wiles's proof, more than three hundred years of struggle had finally come to an end. Fermat's last theorem was a theorem at last.

That was not the end of the story, however. Wiles's new proof showed that "many" elliptic curves are modular, as did his original, faulty proof. But this time, among the elliptic curves for which his proof worked were the ones required to provide Fermat's last theorem. The question remained: Was *every* elliptic curve modular, as the full Shimura-Taniyama conjecture proposed?

Before Wiles's work, most experts had come to the conclusion that the full conjecture was probably true, but no one had the slightest idea how to prove it. Indeed, Ribet believed it would not be proved within his lifetime. So great was Wiles's work, however, that, after his 1994 proof had been acknowledged to be correct, those same experts felt it would not be long before the new methods were strengthened and extended to prove the full conjecture. The hunt was on.

It was to take five years. In the summer of 1999, Brian Conrad and Richard Taylor of Harvard University, Christophe Breuil of the Université de Paris-Sud, and Fred Diamond of Rutgers University announced that they had finally managed to build on Wiles's work to prove the full Shimura-Taniyama conjecture. As this book went to press, mathematicians were still waiting to see a complete proof of the new result, so officially the jury was still out as to the correctness of the claim. But the unofficial word was that it looked okay. Assuming the proof is correct, the new result will mark the culmination of a major advance in mathematics that will form a starting point for a whole new era of research in number theory. For whereas Fermat's last theorem itself was little more than a curiosity from the seventeenth century, having no significant consequences, the Shimura-Taniyama conjecture is a rich and powerful result offering the potential for a great many wide-ranging applications.

Clearly, in number theory, as in many other areas of mathematics, the new golden age is by no means over.

Suggested Further Reading

Few professional mathematicians would be able to follow the recent work of Faltings, Ribet, and Wiles without spending several months acquiring the necessary background knowledge. For the reader of this book who wants to gain a better sense of the earlier work on the last theorem, the necessary background can be found in David Burton's excellent textbook *Elementary Number Theory* (Allyn & Bacon, 1980). A further reference at roughly the same level is the book *Algebraic Number Theory*, by I. N. Stewart and D. O. Tall (Chapman & Hall, 1979). Simon Singh's book *Fermat's Enigma*, published by Walker & Co. in 1997, describes the events that led up to Wiles's proof but does not provide any of the details of the proof itself. (The mathematical level is much lower than in the book you are reading; instead, the focus is on the story and the personalities of the individuals involved.)

The Efficiency of Algorithms

Algorithms Again

The concept of an algorithm has already played a prominent role in this book, in chapter 6. (I shall assume from now on that you have read that chapter.) With Hilbert's tenth problem (considered there), the issue was whether or not a particular problem could be solved by an algorithm, with the stress on the "could be." Pure existence was the order of the day—there was no question about whether the algorithms being discussed were at all practicable. In the context of Hilbert's question, this was, of course, in order. But for problems arising in the real world around us, the mere existence of an algorithm is by no means the end of the matter. Indeed, it is just the start, as there is no real advantage to be gained from having an algorithm that, although theoretically capable of solving the problem at hand, might require thousands of years of high-speed computer time to do so. For the kinds of problems that are of concern to business-persons and applied scientists, the important issue is the existence of an efficient algorithm. In a business application, this might mean the ability to obtain a solution within a few hours. For something like an aircraft-guidance system, results need to be available within a fraction of a second. For applications such as these, it clearly is useful to be able to prove that a particular problem can or cannot be solved by means of an "efficient" algorithm. To do this, the first step is to formulate a suitable method for evaluating an algorithm's efficiency.

The speed with which a given task can be performed on a computer depends on a number of factors. The size and operating speed of the computer, the efficiency of the programming language used to write the program, and the skill of the programmer all are relevant. But these are highly specific factors not suited for a general study. What we require is a very general way of dividing algorithms into two categories: efficient and nonefficient. This classification should be sufficiently robust that altering such "peripheral" factors as the computer speed or the programming language does not convert an inefficient algorithm to an efficient one, or vice versa.

Such an efficiency classification was introduced by A. Cobham and J. Edwards in the mid-1960s and now forms the basis for most work on algorithm efficiency. Although their fundamental measure of efficiency is referred to in terms of "time," in order to avoid a dependence on computing speeds, the actual definition is given in terms of the number of steps required to perform the calculation. Of course, even this notion is not absolute, as it depends on what constitutes a basic step and on how the data is represented. But it turns out that all such considerations are irrelevant to the fundamental notion of efficiency. For this reason, it has become standard to formulate the various definitions in terms of Turing machines (see chapter 6). These are simple enough to allow for a smooth mathematical theory, and yet all the results obtained are equally valid if you recast everything in terms of your favorite computing device, whatever it may be.

Having decided on the Turing machine as the basic computing device for the theory, the idea is to measure the efficiency of an algorithm by the number of steps (i.e., Turing-machine steps) taken to complete the calculation. Questions about how the algorithm is coded as a Turing-machine program and how the data is coded on the tape turn out to be of no significance (i.e., such considerations do not affect the boundary between efficient and inefficient algorithms). What is relevant is the size of the data that must be handled. The more data there is, the more steps are required to handle it. For example, if you are multiplying pairs of integers by hand, it will take you a bit more than four times as long when you move up to numbers twice the length of the previous pair, because there are four times as many basic digit multiplications, plus the associated "overheads" of keeping track of the various carries. Bearing this in mind, here are the basic definitions.

An algorithm (a Turing-machine program for the purpose of this definition) is said to run in *polynomial time* if there are fixed integers A and k such that for input data of length n, the computation is completed in at most An^k steps (for any value of n).

For example, the standard algorithm for adding (by hand) two whole numbers runs in polynomial time. If the numbers are expressed in standard decimal format and the basic computational operation is the addition of two digits, then the addition of two numbers each with $n/2$ digits (input data of length n) will take exactly n steps (allowing for carries), so the preceding definition is satisfied with both A and k equal to 1. In the multiplication of two $n/2$-digit numbers, there are $n^2/4$ basic digit multiplications plus $n/2$ carries, giving $n^2/4 + n/2$ steps in all. Since $n^2/4 + n/2$ is always less than n^2, if you take $A = 1$ and $k = 2$ in the definition, you will see that integer multiplication (by the standard method) is a polynomial-time algorithm.

If the preceding examples are given in terms of Turing machines rather than decimal digit arithmetic, you would need to use larger values of the constant A, and possibly even a larger k as well, but you would still be dealing with a polynomial-time algorithm. In fact this is why the notion of a polynomial-time algorithm is independent of any changes in machine and programming details. These lead only to changes in the size of the two constants; the definition remains valid.

Algorithms that do not run in polynomial time are said to run in *exponential time*. For example, an algorithm that requires 2^n (or 3^n, or n^n, or $n!$) steps to handle input data of length n is an exponential-time algorithm. This explains the use of the word *exponential* here, although the usage is a bit misleading, since it includes functions like $n^{\log n}$, which is not usually regarded as an "exponential" function.

As you will have realized by now, "efficient" algorithms are those that run in polynomial time—and "inefficient" ones are those that require exponential time. The discussion of exponential growth in chapter 1 should be enough to convince you that exponential-time algorithms might be highly inefficient (but see later), but you might well be skeptical about whether polynomial-time algorithms are necessarily efficient. The freedom to choose the constants A and k in the definition of polynomial time would seem to provide far too much leeway; an algorithm that is classified as "efficient" only when you choose $A = 10^{10}$ and $k = 100$ is hardly likely to be "efficient" in any real sense. Two points are important here. First, what happens in practice is that problems can be solved either by exponential-time algorithms only or by polynomial-time algorithms of the order of, say, $10n^3$ steps, or perhaps even fewer. Second, the polynomial/exponential split is only a crude preliminary classification. In the future, it might be necessary to look for a more sensitive distinction, but at the moment the one available works quite well. Both these remarks are emphasized to great effect in table 4.

To illustrate the way the preceding notions are used to classify real problems and algorithms, we now consider a famous problem whose importance in the commercial and business world is self-evident.

The Traveling-Salesman Problem

Imagine you are a traveling salesman and that you must visit a number, say 50, of different locations. The order in which you make your visits is not important as long as you get to them all. It is obviously in the interest of both yourself and whoever is paying your transport costs that you make your visits in the order that requires the least amount of traveling (in total). How do you choose your route? Very likely you start by drawing up a table showing the distance between each pair of locations to be visited. But having done that, then what do you do? For example, what is the most economical route for visiting all the locations in figure 66?

One approach is to list all possible routes, work out the total distance for each, and pick the one that gives the smallest answer. This certainly will work, which demonstrates that the problem can be solved by means of an algorithm, since the procedure could easily be performed by a computer, at least in principle. But even for quite modest numbers of locations, the number of possible routes is far too great to handle. If there are N locations to visit, then there are $N!$ possible itineraries. (Recall that $N!$, read as "N-factorial," is the product of all the numbers N, $N-1$, $N-2$, down to 3, 2, and 1.) Since the function $N!$ is certainly exponential (it grows faster than 2^N or 3^N, though not as fast as the "superexponential" function N^N), listing all possible routes will obviously lead to an exponential-time algorithm. To see just how badly this method will fare, note that for 10 locations, there are

TABLE 4 Polynomial and exponential running times. Assume that a computing device performs one basic operation in 0.000001 seconds. For a given size of data and "time-complexity function" (i.e., how the computation depends on the data size), the table gives the time required to perform the computation. Notice the much more explosive growth rates for the two exponential functions. The computation time for $n = 50$ and a time-complexity function 3^n exceeds the best current estimates of the age of the universe, and for $n = 60$, it is about 100,000 times longer even than that!

Time-complexity function	Size of data: n					
	10	20	30	40	50	60
n	0.00001 sec	0.00002 sec	0.00003 sec	0.00004 sec	0.00005 sec	0.00006 sec
n^2	0.0001 sec	0.0004 sec	0.0009 sec	0.0016 sec	0.0025 sec	0.0036 sec
n^3	0.001 sec	0.008 sec	0.027 sec	0.064 sec	0.125 sec	0.216 sec
2^n	0.001 sec	1.0 sec	17.19 min	12.7 days	35.7 years	366 centuries
3^n	0.059 sec	58 min	6.5 years	3855 centuries	2×10^8 centuries	1.3×10^{33} centuries

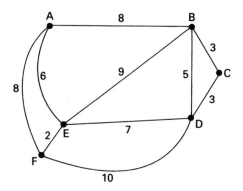

<small>FIGURE 66</small>   The traveling-salesman problem. Find a tour of all the locations shown that minimizes the total distance traveled. (The location-to-location distances along the available paths are indicated.) For instance, the tour ABEFDC means a journey of length $8 + 9 + 2 + 10 + 3 = 32$. (Sometimes the tour must start and finish at the same location, in which case this particular route would be incomplete.)

$$10! = 3,628,800$$

possible routes. This could be handled on a modern computer, but when we get up to 25 locations, the number of routes to consider is a daunting 16 followed by 25 zeros. Furthermore, a tour of 25 cities is not at all unrealistic for a real-life salesman, to say nothing of other situations involving essentially the same mathematical problem in which the number of "locations" might be in the hundreds.

So arriving at a solution by listing all the possibilities is out of the question except when only a small number of locations are involved. What else might you try? A "commonsense" solution maybe? For instance, by looking at the map (or table of distances), you could work out a route that has you first visiting all the locations close to your starting point and then moving farther afield. Although this strategy (and any other you might like to try) may work in certain situations, it has been shown that it will not work in all cases, and it is the overall behavior of the algorithm that concerns us now. Individual instances of the salesman's problem might turn out to have simple solutions (e.g., if all the locations lie on a straight line, the shortest route is obvious), but we want an algorithm that will work in all cases. However, despite all the effort that has been put into the problem (several hundred technical papers have been written on it) since it was first raised by the Viennese mathematician Karl Menger in 1930, the traveling-salesman problem has resisted all attempts at a general solution. In fact, as we shall see, the evidence is strong that there is no efficient algorithm for this problem.

In the meantime, I should mention the considerable advances that have been achieved for some special cases. For a start, in 1962 Michael Held and Richard Karp of IBM used a technique called *dynamic programming* to solve the problem for all tours of at most 13 locations (13! = 6,227,020,800). In 1963, James Little, U. S. R. Murty, Dennis Sweeney, and Richard Karp invented a powerful technique called *branch and bound* that enabled the problem to be solved for tours of up to 40 locations in just a few minutes of computer time on a mainframe computer (an IBM 7090). In 1970, Held and Karp developed a branch-and-bound algorithm that was able to solve a specific instance of the problem, with 42 locations. This algorithm required the examination of only 61 of the vast number (33 followed by 49 zeros) of possible routes. (The significance of this particular problem, which involved 42 cities spread across the United States, was that a solution had been obtained in 1954 by George Dantzig and colleagues at the Rand Corporation.) And in 1979, Harlan Crowder and Manfred Padberg solved a specific problem involving 318 locations, at the time by far the largest number ever handled. In 1987, Padberg teamed up with Giovani Rinaldi to solve a problem with 2,392 cities, using a new method called *branch and cut* that they developed.

In 1992, David Applegate, Robert Bixby, William Cook, and Vasek Chvatal upped the ante by solving a problem involving 3,038 cities. Their solution used a new algorithm based on a technique called *cutting planes,* which steadily improves the collection of linear inequalities describing the set of feasible solutions. In 1993, the same researchers used their new technique to solve a problem with 4,461 cities.

More recently, in 1995, the Applegate team not only pulled out all the stops but also pulled a potpourri of ideas out of the hat to solve a problem with 7,397 cities. Commenting on their work, Bixby said that, "One the most exciting aspects of this work was the fact that it was truly interdisciplinary. It involves ideas from polyhedral combinatorics and combinatorial optimization, integer and linear programming, computer science data structures and algorithms, parallel computing, software engineering, numerical analysis, graph theory, and more."

The Applegate team have since improved their algorithm enormously, and in 1998 used it to solve the traveling-salesman problem for all 13,509 cities of the United States having populations of 500 or more. That computation took three and a half months using three high-performance, multiprocessor servers and an interconnected cluster of 32 Pentium PCs.

If you don't demand an exact solution to the traveling-salesman problem, but only one within a stipulated percentage of the optimal solution, then computation times come down dramatically. For example, in 1991, Jon Bentley developed an algorithm that could find a good approximate solution to any traveling salesman problem involving 10,000 or fewer cities in less than an hour on a desktop workstation.

For most practical applications, a good approximate answer is as good as an accurate solution. Indeed, if the approximate answer can be found in minutes and it takes weeks or months to find the optimal solution, the approximate answer is surely preferable. This is just as well, for as we shall see next, there is evidence to suggest that the traveling-salesman problem is one of a number of computational tasks that to all intents and purposes cannot be solved. More precisely, the computational effort required to find an (accurate) solution increases dramatically as the size of the problem goes up—the very same phenomenon that we met in Chapter 1, where we discussed how the practical impossibility of factoring large whole numbers provides a basis for a secure public key cryptosystem.

P and NP

For an abstract discussion of how amenable problems are to solution using efficient algorithms, it is convenient to reformulate all problems so that they require only a simple yes or no answer. (This allows different problems to be compared.) For example, the multiplication problem (Given integers a and b, what is their product?) could be recast as, Given integers a, b, and c, does $ab = c$? The traveling-salesman problem could be recast as, Given a collection of locations, plus a table of the distances between them, and given a number B, is there a tour of the locations whose total traveling distance is at most B? (It is not immediately apparent that this entirely captures the essence of the problem as originally formulated, but in fact it does. If there is an efficient algorithm for solving the original version, then there is one for solving the simplified version, and vice versa.)

Problems for which yes or no answers are required are called *decision problems*. A decision problem is said to be of type P if it can be solved by an algorithm that runs in polynomial time. For example, the multiplication problem cited earlier is of type P. To see if $ab = c$, simply multiply a and b together and see if the result is equal to c. This requires only polynomial (in fact, quadratic) time.

The traveling-salesman problem may or may not be of type P; at the moment, it is not known for certain. What is known is that the problem is of type NP, which stands for *nondeterministic polynomial time*. To understand this notion, imagine a Turing machine (or any other computation device) that is capable of making random guesses at various stages during its operation. (This must be imagined because there is no possibility of building such a machine.) By using such a hypothetical device (it is called a *nondeterministic Turing machine*), the traveling-salesman problem can be solved in polynomial time. The algorithm is simple. Guess the first location to be visited, then the second, then the third, and so on, until the entire tour has been guessed. Calculate the total distance of the tour and compare this with the given number B. Provided the machine "guesses right" at each stage (in reality, an unlikely event having probability $1/N!$, where N is the number of locations to be visited), the result obtained will be correct. This is what it means for a problem to be of type NP; by making one or more "correct" (or "optimal") guesses, it is possible for a nondeterministic Turing machine to solve the problem in polynomial time.

Another problem of type NP is testing integers to see if they are composite (i.e., not prime). To test a given integer n, guess integers a and b less than n and check if $ab = n$. This clearly gives an answer in polynomial time, and an optimal guess gives the correct answer. But notice that the same approach is not sufficient to establish that the "complementary problem" of determining if a given integer n is prime is of type NP. A single good guess is all that is required to show that n is composite, whereas all guesses must fail in order to prove that n is prime. In fact, primality testing *is* of type NP, but to demonstrate this, you must use a different method of primality testing.

The importance of this highly abstract concept of NP-type problems comes from a combination of two factors. First, many of the problems for which no efficient algorithm has yet been discovered turn out to be of type NP. (Intuitively, we can see that this is because the difficulty in such problems arises not from the necessary computational procedure but solely from the vast number of possibilities. When all these different possibilities are sufficiently alike to be handled in the same manner, they are amenable to the guessing strategy that lies behind the NP concept.) Thus NP provides a theoretical framework for handling many real, practical problems.

The second factor stems from Stephen Cook's 1971 work on algorithm efficiency. By using techniques dating back to Turing and others, Cook was able to provide a means of demonstrating that certain NP-type problems are highly unlikely to be solvable by an efficient, polynomial-time algorithm. (The phrase "highly unlikely to be solvable" might be replaceable by "definitely not solvable," as we shall see.) Specifically, what Cook did was to prove that a particular problem of type NP was what he called *NP complete*. This means that if this particular problem can be solved by a polynomial-time algorithm, so too can every other problem of type NP. In other words, the problem that Cook considered is "just as hard" as every other problem of type NP. Capitalizing on Cook's result, other mathematicians subsequently proved that many other NP problems also are NP complete—including the traveling-salesman problem (see box C).

As a result of the work by Cook and others, there is a way of demonstrating that many of the (type NP) problems that arise in the real world are as hard to solve as is any other problem of type NP. Most mathematicians would conclude that it is a waste of time trying to find an efficient (i.e., polynomial-time) algorithm to solve a problem known to be as hard as every other NP problem. Hence a proof that a given problem is NP complete is inevitably regarded as proof that it cannot be solved by a polynomial-time algorithm.

But there is a difficulty here. Cook's result (and all the subsequent ones to date) does not preclude the possibility that the classes P and NP

Box C

Some NP-Complete Problems

The Traveling-Salesman Problem
(See the text for details.)

The Hamiltonian Circuit Problem
Given a network of cities and roads linking them, is there a route that starts and finishes at the same city and visits every other city exactly once?

The Multiprocessor Scheduling Problem
Given a collection T of tasks to be performed, together with a list of the times taken to perform each task on a certain type of processor, and given also a specific number of processors of that type, is it possible to split up the tasks in T and assign each group to one processor so that the total time taken to perform all the tasks is less than some specified time? (Each processor works sequentially, although the entire collection of processors runs concurrently.)

The Map-Coloring Problem
(See chapter 7 for the background.) Is it possible to color a given map using only three colors in such a way that no two countries with a common border are colored the same?

The Quadratic Residue Problem
Given positive integers a, b, c, with a less than b, is there a positive integer x less than c such that

$$x^2 \bmod b = a?$$

Quadratic Diophantine Equations
(See chapter 6 for the background.) Given positive integers a, b, c, do there exist positive integers x and y such that

$$ax^2 + by = c?$$

are in fact the same; that is, that any problem of type NP can in fact be solved using a polynomial-time algorithm (although finding such an algorithm might be no easy matter in any specific instance). And if this is the case, knowing that a problem is "as hard as any other in the class NP" does not really say much. (All NP problems would be "easy" in the current sense.) But few experts regard this as a likely possibility. The nature of the NP concept, involving as it does the highly nonalgorithmic procedure of "guessing" (in fact, "guessing right"—usually against phenomenal odds), makes it improbable that it can be equivalent to type P. Consequently, the theoretical possibility that P and NP are the same is generally discounted at the outset, and an NP-completeness result *is* regarded as proof that the problem is genuinely "unsolvable."

Of course, all that it would require to settle the issue conclusively would be to find just one problem that is of type NP but demonstrably not of type P. But despite the intuitive gap between the two notions P and NP, this has so far not been achieved, and indeed all the available evidence suggests that it is an extremely hard problem. It is known as the "P = NP problem," and it is, as you might imagine, regarded as one of the most significant open problems of present-day computational mathematics. Part of its significance stems from its relevance to many practical problems. But here one must be cautious, for questions of relevance are never simple, as the following section reminds us.

Back to Real Life: Linear Programming

Although they can provide some valuable information, the theoretical techniques just described do not always give a completely accurate picture of whatever it is to which they are applied. An algorithm may in theory run in exponential time (i.e., "inefficiently") and yet, in practice—with everyday data—it may in fact perform very well. The exponential running may arise only for certain kinds of data rarely encountered; to some extent, the traveling-salesman problem falls into this category. When applied to real-life configurations of cities and roads, the available methods—which are without doubt exponential in their running time—can perform quite well. An even more striking example of the potential gulf between theory and practice is provided by the so-called linear-programming problem, the problem that gave rise to (and continues to be one of the central themes of) the subject known as *operations research.* (This subject, whose origins lie in World War II, uses mathematical methods to assist in complex problems involving the direction and management of large systems of workers, machines, materials, and money in industry, business, government, and defense.)

Linear programming is a technique used to provide a mathematical description (or *model*) of a real-life problem in which something needs to be *maximized* (e.g., profits or security) or *minimized* (e.g., costs or risks). The required *optimization* is achieved by a suitable choice of the values of a number of *parameters* (or *variables*). Both the factor to be optimized and some or all the parameters are subject to one or more *constraints.* The word *linear* in linear programming indicates that all the mathematical expressions in the model are linear (i.e., do not involve the multiplication together of two or more variables or the raising of a variable to a power). In practice, this is not a major restriction, since most of the problems encountered in real life either are intrinsically linear or can be assumed to be linear without giving rise to any great errors.

An elementary consideration of the problem shows that linear constraints have a natural geometrical representation. Values of the variables that satisfy all the constraints correspond to points that lie inside a certain geometric figure. If there are two variables, that figure will be a polygon (whose number of sides corresponds to the number of constraints); if there are three variables, it will be a polyhedron; and if there are *N* variables, it will be a polytope in *N*-dimensional space (see chapter 9). Of course, we cannot draw a polytope in four or more dimensions, but the mathematics is straightforward whatever the dimension.

A simple example should make this clear. Imagine a company that makes two kinds of cloth, *A* and *B*, using three different colors of wool. The amounts of wool required to make a unit length of each type of cloth, and the total amount of wool of each color available, are given in table 5. The manufacturer's profit is $12 per unit length on cloth *A* and $8 per unit length on cloth *B*. The question is, How should the available wool be used to make the largest possible overall profit?

We begin by letting *x* and *y* denote the number of units of cloth *A* and cloth *B*, respectively, that are produced. This yields a profit *P* (in dollars) of

$$P = 12x + 8y. \tag{8}$$

TABLE 5   The amounts of red, green, and yellow wool required to make unit lengths of cloths *A* and *B*, and the total amounts available.

Color of Wool	Amount Required per Unit Length		Amount Available
	Cloth A	*Cloth B*	
Red	4 kg	4 kg	1400 kg
Green	6 kg	3 kg	1800 kg
Yellow	2 kg	6 kg	1800 kg

What are the constraints on the values of *x* and *y?* Since only 1400 kg of red wool are available and both types of cloth require 4 kg of red wool for each unit length,

$$4x + 4y \leqslant 1400. \tag{9}$$

Similarly, by considering the amount of available green and yellow wool,

$$6x + 3y \leqslant 1800, \quad 2x + 6y \leqslant 1800. \tag{10}$$

Finally, since neither *x* nor *y* should be negative (a constraint that is obvious when the actual problem is considered, but that must be made explicit in the mathematical representation), there are the requirements that

$$x \geqslant 0, \quad y \geqslant 0. \tag{11}$$

Figure 67 is a graphical representation of the constraints imposed by the inequalities (9), (10), and (11). Any pair of values of *x* and *y* that satisfy all these constraints will be the coordinates of a point within the shaded region, and conversely, any point in this region will have coordinates

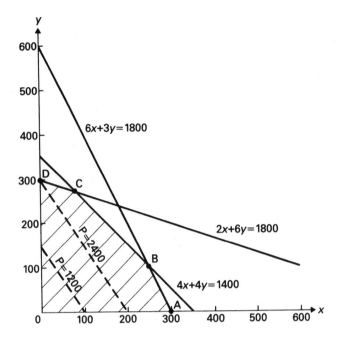

FIGURE 67   Linear programming. The solution to the cloth-manufacturing problem (see the text for details).

that satisfy the inequalities (9), (10), and (11). (Check this yourself by reading the coordinates of various points from inside and outside the region). So, what we must do now is find a point within the shaded area that makes the quantity P in equation (8) as large as possible.

Now, all lines whose equations have the same form as equation (8) for a fixed value of P are parallel to one another. (Two such lines, for $P = 1200$ and $P = 2400$, are shown in figure 67.) It is fairly clear, therefore, what we must do to maximize P: move the profit line—given by equation (8)—as far away from the origin as possible without completely leaving the shaded region. This takes us to the point marked B. The coordinates of B are easily obtained by elementary algebra (as the solution of two simultaneous equations): they are (250, 100). Thus the manufacturer must make 250 units of cloth A and 100 units of cloth B to obtain the maximum possible profit, which is $3800. (As it happens, this will use all the available red and green wool but leave 700 kg of yellow left over, so our manufacturer should have done his homework before buying his wool.)

Having solved the problem, notice what was involved. The constraints were represented in figure 67 as the polygonal region ABCDO of the plane. The maximization point was one of the corners of the polygon—the remaining task was to find which corner. In this simple example, finding which corner presented no difficulty, but this is the part that makes more complicated (and hence more realistic) linear-programming problems hard. In a problem with three variables, the constraints will give rise to a three-dimensional polyhedron; with N variables you get an N-dimensional polytope, which cannot be drawn but which can still be handled algebraically. In every case, the problem boils down to finding the vertex of the constraint figure (polygon, polyhedron, or polytope) where optimization occurs. But how do you do this? There could be millions (or more) corners, so an exhaustive search is usually out of the question, just as with the traveling-salesman problem. You need a systematic way of flushing out the optimal corner.

In 1947, the American mathematician George Dantzig devised such a method: the *simplex algorithm.* In essence, this method starts from one vertex (exactly how this initial vertex is found will not be described here) and then moves around the surface of the polytope, following edges from vertex to vertex. Each time a new vertex is reached, there are several (perhaps many) different ways of proceeding next, and there are various criteria for deciding which one to follow (the most "obvious" of which is to move only to a vertex that increases the quantity to be maximized—or decreases it if it is to be minimized).

Because of the enormous number of possible paths around the edges of a polytope, the simplex algorithm is known to be, in theory, an exponential-time algorithm, but when used in practice (on problems involving hundreds or even thousands of variables), it works extremely well, homing in on the optimal vertex in a relatively small number of steps. Indeed, the indications are that it tends to run in linear time (i.e., the number of steps is directly proportional to the number of variables involved). The types of constraints (and their associated polytopes) that cause the algorithm to operate inefficiently just do not arise very often in practice—they must be specially "cooked up" with the express purpose of defeating the simplex method. Indeed, they are so contrived that the existence of these "artificial" problems does not prevent professionally implemented versions of the simplex algorithm from being one of the first packages made commercially available whenever a new computer system is put on the market for industrial and commercial use. Put simply, the method works.

But is there a faster method, one that is fast not just for most problems but in all cases—a polynomial-time algorithm? Intuitively, instead of finding the optimal polytope vertex by following edges around the surface of the polytope, it should be faster to "take a shortcut" across the interior. The difficulty with this approach is that because you do not know in advance which vertex is optimal, how can you decide in which direction to proceed? By staying on the surface of the polytope, at least you have a method for deciding which way to go at each stage. But is there any way of orienting yourself once you have "cast loose" from the surface and headed into the interior?

There is. In 1970, the Soviet mathematician N. Z. Shor realized that an old technique known as Newton's method could be applied to the linear-programming problem, and further modifications of this idea by G. M. Levin, D. B. Judin, and A. S. Nemirovski (also from the Soviet Union) led to the formulation in 1976 of the so-called ellipsoidal method, in which the direction of the path to be followed across the interior of the polytope is determined with the aid of a sequence of ellipsoids drawn to "approximate" the polytope. In 1979, Khachian (again from the Soviet Union, which had a virtual monopoly on this technique) showed that the ellipsoidal method runs in polynomial time. Unfortunately, although this meant that the method was theoretically better than the simplex method, when applied to real-world problems, it did not perform anything like as well as the simplex algorithm.

So much for the theoreticians' concept of efficiency, you might say, when the theoretically inefficient method easily outperforms the theoretically efficient one. And many nonmathematicians did say just that. After all, the linear-programming problem was—and is—perhaps the single most important real-life mathematical problem. If there were to be any example to indicate a deficiency in the polynomial-time/exponential-time classification of efficiency, this was the worst one possible from the pure mathematicians' point of view.

But then early in 1984, another theoretician came to the rescue. Narendra Karmarkar, a 28-year-old mathematician working for Bell

Laboratories in the United States, discovered a polynomial-time linear-programming algorithm that really did work well in practice, on many occasions even outperforming the simplex method to a remarkable degree. (In one test with a 5000-variable problem, Karmarkar's algorithm was fifty times faster than the simplex algorithm.) It was a remarkable and quite unexpected advance, and to obtain his new algorithm, Karmarkar had to use some highly sophisticated mathematics involving a sequence of "reshapings" of the polytope in order to obtain "preferred directions" to follow once you are inside it. (Though just as in computer implementations of the simplex algorithm, where the computer deals with arithmetic manipulations and not directly with the geometrical ideas behind them, so too when the Karmarkar algorithm is implemented, the sophisticated geometrical ideas are suppressed in favor of a series of arithmetic operations on matrices.)

So the theoreticians' concept of efficiency was shown to work well after all. Moreover, the new algorithm provided a striking example of how some highly sophisticated abstract mathematics, involving multidimensional analogues of polyhedra and bizarre mathematical "deformations," can lead to a concrete product of crucial importance in the "real" world of business, commerce, and defense. Altogether an exemplary blend of the pure and abstract with the world we live in and an excellent place to bring to an end this survey of mathematics' New Golden Age.

Suggested Further Reading

The standard introductory text for the study of algorithm efficiency is *Computers and Intractability*, by Michael Garey and David Johnson (Freeman, 1979).

Another good source is *Computational Complexity*, by Christos Papadimitriou (Addison-Wesley, 1994). For a discussion of the traveling salesman problem, see the article "The Traveling Salesman Problem: A Case Study, by D. S. Johnson and L. A. McGeach, in the collection *Local Search in Combinatorial Optimization*, edited by E. H. L. Aarts and J. K. Lenstra (Wiley, 1997).

For a low-level account of the linear-programming problem and the simplex algorithm, see the book *An Illustrated Guide to Linear Programming,* by Saul Gass (Dover, 1990).

Karmarkar's algorithm is described in his paper "A New Polynomial-Time Algorithm for Linear Programming," published in the mathematical journal *Combinatorica,* no. 4 (1984):373–392.

Index